Ying Ma

Phytoremediation via Hyperaccumulators-Beneficial Bacteria System

Ying Ma

Phytoremediation via Hyperaccumulators-Beneficial Bacteria System

Bacterial Assisted Phytoremediation

LAP LAMBERT Academic Publishing

Impressum / Imprint

Bibliografische Information der Deutschen Nationalbibliothek: Die Deutsche Nationalbibliothek verzeichnet diese Publikation in der Deutschen Nationalbibliografie; detaillierte bibliografische Daten sind im Internet über http://dnb.d-nb.de abrufbar.
Alle in diesem Buch genannten Marken und Produktnamen unterliegen warenzeichen-, marken- oder patentrechtlichem Schutz bzw. sind Warenzeichen oder eingetragene Warenzeichen der jeweiligen Inhaber. Die Wiedergabe von Marken, Produktnamen, Gebrauchsnamen, Handelsnamen, Warenbezeichnungen u.s.w. in diesem Werk berechtigt auch ohne besondere Kennzeichnung nicht zu der Annahme, dass solche Namen im Sinne der Warenzeichen- und Markenschutzgesetzgebung als frei zu betrachten wären und daher von jedermann benutzt werden dürften.

Bibliographic information published by the Deutsche Nationalbibliothek: The Deutsche Nationalbibliothek lists this publication in the Deutsche Nationalbibliografie; detailed bibliographic data are available in the Internet at http://dnb.d-nb.de.
Any brand names and product names mentioned in this book are subject to trademark, brand or patent protection and are trademarks or registered trademarks of their respective holders. The use of brand names, product names, common names, trade names, product descriptions etc. even without a particular marking in this work is in no way to be construed to mean that such names may be regarded as unrestricted in respect of trademark and brand protection legislation and could thus be used by anyone.

Coverbild / Cover image: www.ingimage.com

Verlag / Publisher:
LAP LAMBERT Academic Publishing
ist ein Imprint der / is a trademark of
OmniScriptum GmbH & Co. KG
Heinrich-Böcking-Str. 6-8, 66121 Saarbrücken, Deutschland / Germany
Email: info@lap-publishing.com

Herstellung: siehe letzte Seite /
Printed at: see last page
ISBN: 978-3-659-69505-6

Contents

List of Tables

List of Figures

Preface

Soil contamination with heavy metals is becoming one of the most severe environmental problem, leading to losses in agricultural yield and hazardous health effects as they enter the food chain. Application of biological decontamination techniques is a challenging task because heavy metals cannot be degraded and hence in the environment indefinitely. The established conventional techniques for cleanup of metal contaminated soils are generally very costly and also destructive to the soil. Phytoremediation has been proposed as an alternative method to remove pollutants from air, soil and water or to render pollutants harmless and does not affect soil biological activity, structure and fertility. Unfortunately, even hyperaccumulating plants that are relatively tolerant of various metal contaminants usually remain small under metal stress. To remedy this situation, plant growth-promoting bacteria (PGPB) that facilitate the growth of various plants especially under environmentally stressful conditions, can make the entire phytoremediation process much more efficacious. Currently, the benefits of combining endophytic or rhizosphere bacteria with plants have been successfully used for metal removal from contaminated soils. Although some heavy metals are toxic to plants and their associated microbes at high concentrations, the metal resistant bacteria have been found to occur widely in tissue interiors or rhizosphere of various hyperaccumulators without causing disease symptoms, indicating that they have evolved to be resistant to high concentrations of heavy metals and thereby conferring their host plant higher metal tolerance. Moreover, PGPB can promote plant establishment, growth and development under adverse conditions through various mechanisms including production of plant beneficial substances (e.g. 1-aminocyclopropane-1-carboxylate deaminase, phytohormones and siderophores), solubilization and transformation of mineral nutrients as well as biocontrol of disease through production of antifungal metabolites (Fig. 1).

1

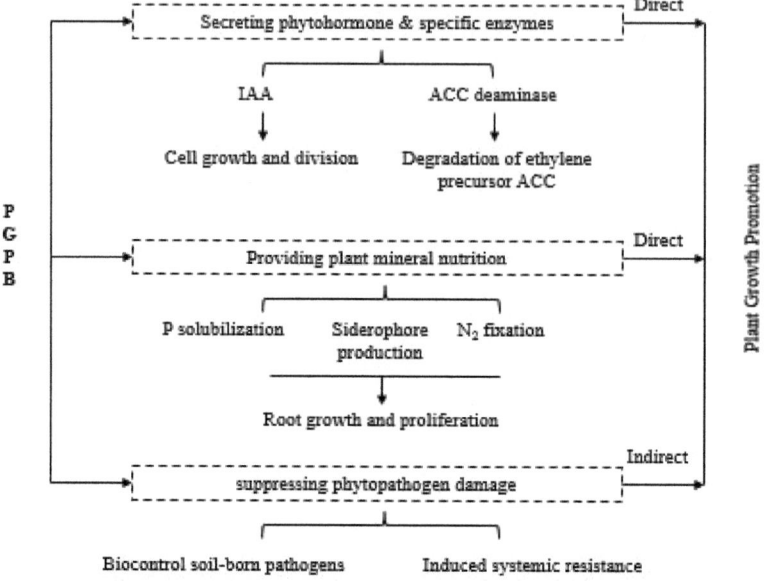

Figure 1 Plant growth promoting mechanisms of PGPB in metal contaminated soils.

In addition, certain endophytic bacteria can potentially reduce phytotoxic effects and alter heavy metal bioavailability by producing siderophores, organic acids, biosurfactants and extracellular polymeric substances (Fig. 2).

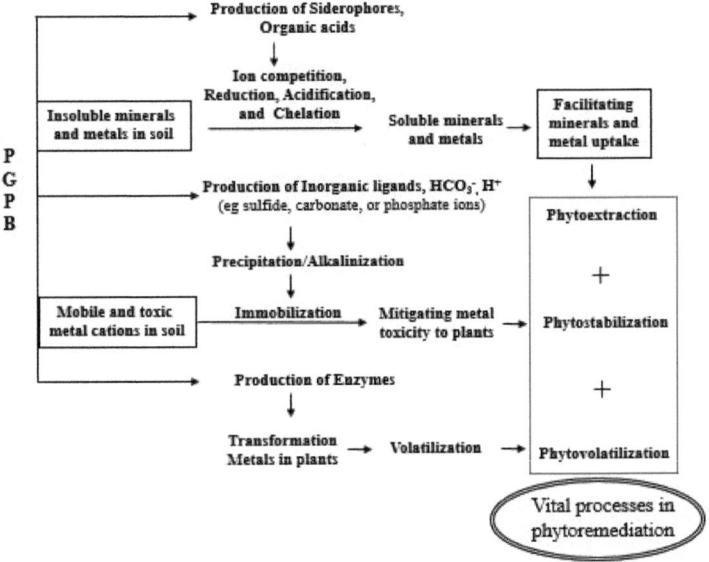

Figure 2 Effects of PGPB on bioavailability of heavy metals in soil.

In spite of these beneficial effects on plant, the PGPB must be able to survive and colonize tissue interior of host plants without causing disease, therefore the colonization potential of PGPB in plants has been considered as a major factor affecting the inoculum efficiency for microbe-assisted phytoremediation. Overall the activities of PGPB enhance the efficiency of phytoremediation processes in metal contaminated soils indirectly by conferring plant metal tolerance and enhancing plant biomass yield in order to remove/arrest the pollutants and directly by enhancing metal mobilization, translocation and uptake (facilitate phytoextraction) or reducing the mobility/availability of metal contaminants in the rhizosphere (phytostabilization). Recent advances have been explored in understanding plant-microbes-heavy metals interactions and their application for metal phytoremediation. The monograph aims to provide a comprehensive overview of various promising phytoremediation processes mediated by PGPB and to illustrate the mechanisms underlying the ability of PGPB to influence heavy metal uptake through various biogeochemical processes including chelation, mobilization, translocation, transformation, volatilization, immobilization, precipitation of heavy metals ultimately facilitating phytoremediation. In this book, I highlight the biodiversity and plant growth promoting traits of metal resistant PGPB and discuss their essential roles in phytoremediation of metal contaminated soils and prospects of their application in practice and orientation of the research in future.

Chapter I Feasibility of enhanced phytoextraction of metal contaminated soils using *Sedum plumbizincicola* and *Phyllobacterium myrsinacearum* RC6b

The aim of this study was to investigate the effects of metal mobilizing plant-growth beneficial bacterium *Phyllobacterium myrsinacearum* RC6b on plant growth and Cd, Zn and Pb uptake by *Sedum plumbizincicola* under laboratory conditions. Among a collection of metal-resistant bacteria, *P. myrsinacearum* RC6b was specifically chosen as a most favorable metal mobilizer based on its capability of mobilizing high concentrations of Cd, Zn and Pb in soils. *P. myrsinacearum* RC6b exhibited a high degree of resistance to Cd (350 mg L^{-1}), Zn (1000 mg L^{-1}) and Pb (1200 mg L^{-1}). Furthermore, *P. myrsinacearum* RC6b showed multiple plant growth beneficial features including the production of 1-aminocyclopropane-1-carboxylic acid deaminase, indole-3-acetic acid, siderophore and solubilization of insoluble phosphate. Inoculation of *P. myrsinacearum* RC6b significantly increased *S. plumbizincicola* growth and organ metal concentrations except Pb, which concentration was lower in root and stem of inoculated plants. The results suggest that the metal mobilizing *P. myrsinacearum* RC6b could be used as an effective inoculant for the improvement of phytoremediation in multi-metal polluted soils.

1. Introduction

Contamination of soils with toxic heavy metals through mining operations, discharge of industrial effluents, extensive use of pesticides and fertilizers etc., is of great concern due to its detrimental effects on soil biological systems (Giller et al., 1998). In China, thousands of abandoned or operating metal based ore mines exist on public lands, which have generated around 1,500,000 ha of metal polluted soil and increases at a rate of 46,700 ha yr^{-1} (MEPPRC, 2006). Results from recent studies (Kachenko and Singh, 2006; Zhuang et al., 2009) also demonstrate that the food crops grown in metal contaminated soils pose a major health concern. For instance, Li et al. (2006) reported that Chinese cabbage and *Brassica napus* grown in the vicinity of a Chinese Pb/Zn mine had higher levels of heavy metals than the maximum permissible value in food proposed by food regulation. Thus, the development of effective remediation strategies for metal polluted soils that do not affect soil biological and ecological health is necessary. Conventional remediation methods such as soil washing and excavation, landfilling of the top contaminated soils, electrokinetic treatment, leaching and immobilization are expensive, time consuming and often harmful to soil biological system. Phytoremediation is a low cost and environmentally friendly technology, which uses plants and their associated microbes for inactivation or removal pollutants from the soil, water, sediments and air (Glick, 2003).

Although numerous plant species are capable of hyperaccumulating specific heavy metal, these plants are not suitable for treating soils contaminated with multiple metals because of their slow growth, low tolerance to multiple metal stress and inability to uptake multiple metals (Ghosh et al., 2011). Recently, microbial mediated plant stress amelioration has emerged as an important component of metal stress management in plants and their role in improving plant growth and phytoremediation process in metal polluted soils has been well established (Rajkumar et al., 2013b). It has been demonstrated that the inoculation of plants with metal-resistant plant growth-promoting rhizobacteria (PGPR) play an important role in improving the efficiency of heavy metal phytoremediation (Ma et al., 2011a; Rajkumar et al., 2012). PGPR such as *Azospirillum*, *Azotobacter*, *Achromobacter*, *Bacillus*, *Burkholderia*, *Gluconacetobacter*, *Pseudomonas* and *Serratia*, have been known to improve plant growth through various mechanisms like production of phytohormones, siderophores and 1-aminocyclopropane-1-carboxylic acid (ACC) deaminase, and solubilization of mineral nutrients (Rajkumar et al., 2008; Sheng et al., 2008; Ma et al., 2009).

Several factors including soil nutrients, pH, plant type and their associated microbial flora etc., affect plant-microbe interactions and thereby influence heavy metal uptake by plants. However, the bioavailability of heavy metals in rhizosphere soils is considered to be an important factor that determines the efficiency of phytoextraction process. Metal tolerant microbes have been frequently reported in the rhizosphere of hyperaccumulators growing in metal polluted soils indicating that these microbes have evolved a heavy metal-tolerance and that they may play significant roles in mobilization or immobilization of heavy metals by excreting various metabolites including organic acids or extracellular polymeric substances (Rajkumar et al., 2012; Sessitsch et al., 2013; Prapagdee et al., 2013). *Sedum plumbizincicola* is one of the hyperaccumulators (Jiang et al., 2010) which has a remarkable capacity to withstand the metal stress in polluted soils and recent experiments have also demonstrated its potential for heavy metal phytoextraction (Wu et al., 2008). Although there is much interest in increasing the phytoremediation efficiency of *S. plumbizincicola*, effects of interactions of metal mobilizing microbes and *S. plumbizincicola* on the heavy metal phytoremediation, to our knowledge, has not been investigated. The objectives of our study were to isolate and characterize metal mobilizing PGPR from the rhizosphere of *S. plumbizincicola* and to investigate the effects of metal mobilizing PGPR on plant growth and Cd, Zn and Pb uptake by *S. plumbizincicola* in multi-metal contaminated soils.

2. Materials and methods

2.1. Isolation of metal tolerant bacterial strain

The bacterial strains were isolated from rhizosphere of *S. plumbizincicola* grown on Pb/Zn mine spoils in Chunan city of Zhejiang, Southeast of China. The physicochemical properties of soil were determined according to standard methods. The selected characteristics of the soil were: pH (1:1 w/v water) 7.6; organic matter 1.36%; copper 1.83 g kg^{-1}; zinc 0.992 g kg^{-1}; cadmium 0.0913 g kg^{-1}; lead 14.2 g kg^{-1}. About 1 g of soil samples were serially diluted using 25 mM phosphate buffer and spread over on sucrose minimal salts low-phosphate (SLP) medium amended with 100 mg L^{-1} of Cd (CdCl$_2$), Zn (ZnSO$_4$), or Pb (Pb(NO$_3$)$_2$). This medium was designed to avoid metal salt precipitation (Sheng et al., 2008). The plates were incubated at 37 °C for 48 h. From the metal-resistant colonies, different strains were picked and purified on the SLP agar medium containing 100 mg metal L^{-1}. In order to isolate an effective metal mobilizing bacterial strain, the metal resistant isolates were tested for their ability to increase the water soluble Cd, Zn and Pb concentrations in soils. The metal contaminated soils were collected from Fuyang city of Zhejiang province, PR China, sieved and sterilized by steaming (100 °C for 1 h on three consecutive days) (Table 1).

Table 1 The physiochemical properties of soils used in metal mobilization and pot experiments.

Soil property	Value
pH (H₂O)	8.1
Cation exchange capacity (cmol kg⁻¹)	11.4 ±0.1
Organic matter (g kg⁻¹)	36.3 ±1.2
Total element concentration	
N (g kg⁻¹)	1.7 ±0.0
P (g kg⁻¹)	1.1 ±0.1
K (g kg⁻¹)	18.6 ±0.2
Cd (mg kg⁻¹)	5.9 ±0.3
Zn (mg kg⁻¹)	736 ±13
Pb (mg kg⁻¹)	153 ±8
Extractable element concentration (1 M NH₄NO₃)	
N (mg kg⁻¹)	107 ±1
P (mg kg⁻¹)	9.4 ±1.3
K (mg kg⁻¹)	60.7 ±0.8

Values represent means ± SD (n=5).

The metal resistant strains were grown in Luria-Bertani (LB) broth and placed on a shaker at 200 rpm and 27 ℃. After 24 h, optical density (OD) was measured at 600 nm and adjusted to 1.5; the cultures were centrifuged at 6000 rpm for 10 min, washed in phosphate buffer (pH 7.0) and resuspended in sterile water. The bacterial cultures (up to 1 mL) were added to the 1 g of sterile metal contaminated soils in the centrifuge tubes. Sterile water was used as the control. All tubes were weighed, wrapped in brown paper and kept on an orbital shaker at 200 rpm and 27 ℃ . The tubes were again weighed after 7 d to compensate for water-evaporation. To extract the soil water-soluble metal, 10 mL of sterile water was added to each tube (Rajkumar et al., 2008). The soil suspensions were centrifuged at 7000 rpm for 10 min and filtered. The concentrations of metals in the filtrate were determined using an atomic absorption spectrophotometer.

2.2. Characterization of metal mobilizing strain

2.2.1. Genetic characterization

The bacterial strain was grown in LB broth at 30 °C and total DNA was extracted using the QuickExtract bacterial DNA extraction kit. The 16S rRNA was amplified using the following primers FAM27f (5′-GAGTTTGATCMTGGCTCAG-3′) and 1492r (5′-GGYTACCTTGTTACGACTT-3′). Each amplification mixture (5 µL) was analyzed by agarose gel (1.5%, w/v) electrophoresis in TAE buffer (0.04 M Tris acetate, 0.001 M EDTA) containing 1 µg mL^{-1} (w/v) ethidium bromide. Partial nucleotide sequence of the amplified 16S rDNA was determined using automated DNA sequencer, and then compared to similar sequences in the GenBank using the BLAST analysis.

2.2.2. Heavy metal resistance levels

To check the metal resistant levels, the selected bacterial strain was grown in LB agar media containing different concentrations of Cd, Zn or Pb ranging from 100 to 1200 mg L^{-1}. Cultures were incubated at 27 °C for 7 d. The highest concentration of metal supporting growth was defined as the maximum resistance level. Moreover, the growth pattern of the isolated bacterial strain in metal contaminated liquid medium was also determined. Briefly, the 250 mL culture flask containing 20 mL LB broth supplemented with heavy metals at the concentration of 200 mg L^{-1} (Cd, Zn or Pb) were inoculated with logarithmic-phase bacterial isolate. All the cultures including controls (in triplicate) were incubated at 27 °C for 36 h at 200 rpm. The bacterial growth was measured once in every 4 h by measuring the OD at 600 nm.

2.2.3. Characterization of plant growth promoting features

The metal mobilizing isolate was screened for the ability to grow on Dworkin-Foster (DF) salts minimal medium (Dworkin and Foster, 1958) with ACC as the sole nitrogen source. The inoculated DF salt minimal medium without ACC was used as a blank. The bacterial growth was monitored by measuring the OD at 600 nm. Further, the ACC deaminase activity was determined as described by Ma et al. (2009). Siderophore production by metal mobilizing strain was detected by the method of Schwyn and Neilands (1987) using chrome azurol S (CAS) agar. The diameters of orange halo produced by the colony on blue agar were indicative of the siderophore biosynthesis

9

level. The presence of catechol and hydroxamate siderophores in iron-restricted bacterial culture supernatants was also quantitatively determined by the calorimetric assay of Arnow (1937) and Atkin et al. (1970) method, respectively.

The metal mobilizing isolate was further analyzed for its ability to solubilize insoluble P using modified Pikovskayas medium (Sundara-Rao and Sinha, 1963). The soluble phosphate in the culture supernatant was quantified as described by Park et al. (2011). Production of indole-3-acetic acid (IAA) by metal mobilizing isolate was assayed as described by Bric et al. (1991) using LB medium with different concentrations of L-tryptophan (0, 1, 2, 3, 4 and 5 mg mL^{-1}).

2.3. Pot experiment

For pot experiments, the soils samples collected from Fuyang city of Zhejiang Province, PR China were dried and passed through a 2 mm sieve (Table 1). The plants, *S. plumbizincicola* were obtained from an old Pb/Zn mine in Chunan city of Zhejiang province, China. The fresh shoot samples (approximately 5 cm long with a pair of leaves and 4-5 nodes) were cleaned with tap water and grown in a half-strength Hoagland's nutrient solution for 7 d. Roots of precultured seedlings were surface-sterilized by sequential immersion in 70% (v/v) ethanol for 1 min, and 3% NaClO for 3 min and washed several times with sterile water. For inoculation of the seedlings, the overnight grown bacterial culture was centrifuged at 6000 rpm for 10 min and the pellet was washed twice with biological saline (0.85% KCl). The pellet was resuspended in biological saline and the OD$_{600}$ was adjusted to 1.5. The roots of seedlings were soaked for 2 h in the bacterial culture or sterile water (controls) and transplanted in plastic pot containing 750 g of metal polluted soil (six plants pot^{-1}). The plant seedlings were allowed to grow in a greenhouse at 25 ± 5 °C and a 16:8 d/night regime. Each treatment was performed in five replicates. After 75 d, the plants were carefully removed from the pots and the root surface was cleaned several times with distilled water. Plant root and shoot length, fresh and dry weight were measured, respectively. The accumulation of metals (Cd, Zn, Pb) in root and shoot system was quantified as described by Ma et al. (2009).

2.4. Statistical analysis

Analysis of variance followed by post-hoc Fisher Least Significant Difference test ($p < 0.05$) were used to compare treatment means. All the statistical analyses were carried out using SPSS 10.0.

3. Results and discussion

3.1. Isolation of metal mobilizing bacteria

Effective microbe-assisted phytoextraction depends on the identification of metal resistant PGPR capable of improving the plant growth and bioavailability of heavy metals in soils and the selection of suitable plants with potential to tolerate and uptake high concentrations of heavy metals. It has been previously reported by several authors that the inoculation of plants with PGPR, could improve the plant survival in metal polluted soils due to the microbial activity/action in the rhizosphere soils (Prapagdee et al., 2013; Srivastava et al., 2013). In particular, the efficiency of heavy metal extraction by hyperaccumulators can be enhanced by inoculating metal mobilizing PGPR. In this study, the metal mobilizing bacteria were isolated from the rhizosphere of *S. plumbizincicola* grown on Pb/Zn mine spoils with an objective to assess the interactive effects of *S. plumbizincicola* and metal mobilizing bacteria on heavy metal phytoremediation. During the initial screening process, a total of 45 morphologically different metal-resistant bacterial strains were isolated. Out of the 45 isolates, strain RC6b was specifically chosen for further studies due to its high metal solubilization ability in soil (Fig. 3).

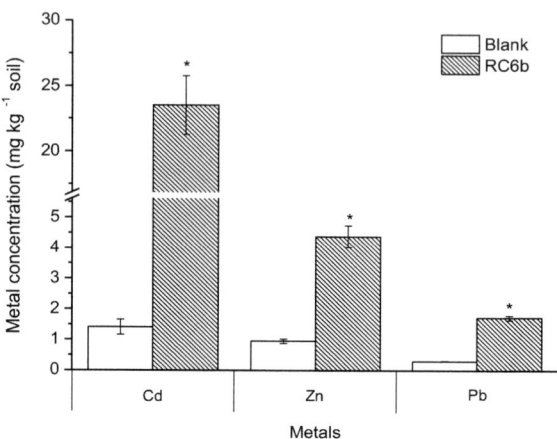

Figure 3 Effect of inoculation with *P. myrsinacearum* RC6b on the solubilization of Cd, Zn and Pb in soil. Bars represent standard deviations of three replicates. An asterisk (*) denotes a value significantly greater than the corresponding control value according to Fisher's protected LSD test ($p < 0.05$).

Compared with non-inoculated control treatment, inoculation of RC6b for 7 d, significantly increased the concentrations of water soluble Cd, Zn and Pb in soil by 16.7-, 4.6- and 5.7-fold, respectively. These results are consistent with those of Rajkumar et al. (2008), Jiang et al. (2008) and Ma et al. (2009), they found an increase in metal concentrations (Ni, Cu, Zn, Cd and Pb) in water soluble fractions in the presence of metal mobilizing bacteria. The observed increase in the concentrations of water soluble metals in this study could be attributed to the effects of microbial metabolites/actions such as altering soil pH, release of organic acids, siderophores and oxidation/reduction reactions (Rajkumar et al., 2012; Rajkumar et al., 2013b).

3.2. Characterization of metal mobilizing RC6b

3.2.1. Genetic characterization

The bacterial strain that showed the highest metal solubilization capacity, RC6b was identified as *P. myrsinacearum* based on the highest sequence similarity (99%) and phylogeny analysis. The obtained sequence (1359 bp) was deposited in the GenBank with accession number JX512224. Phylogenetic tree in Fig. 4 based on 16S rRNA sequence revealed a relationship between isolated strain in this research and other related bacteria reported in the literature.

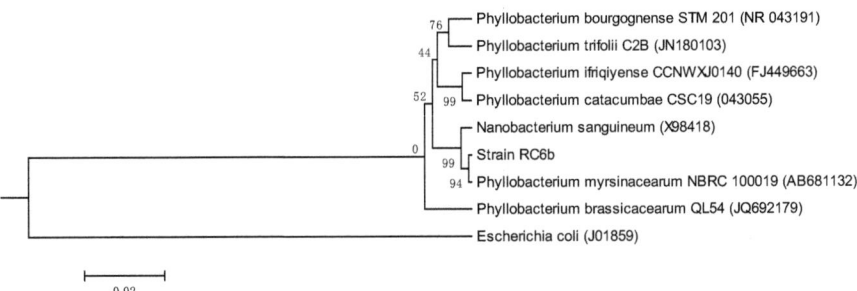

Figure 4 Phylogenetic tree showing the relationship of partial 16S rDNA gene sequences from metal resistant PGPR RC6b with other related sequences from identified bacteria in the database [*P. myrsinacearum* (AB681132), *P. bourgognense* (NR_043191), *P. brassicacearum* (JQ692179), *P. ifriqiyense* (FJ449663), *P. trifolii* (JN180103), *P. catacumbae* (NR_043055), *Nanobacterium sanguineum* (X98418) and *Escherichia coli* (J01859)]. *E. coli* was used as the out-group. The bar represents 0.02 substitutions per site.

3.2.2. Heavy metal resistance levels

The microorganisms isolated from metal contaminated natural environment can be constitutively or adaptatively resistant to increasing metal concentrations and various strategies including physical sequestration, exclusion, complexation and detoxification can be developed by adapted strain to resist high metal concentrations (Nies, 2003). In this study, the strain *P. myrsinacearum* RC6b was found to exhibit multiple heavy metal resistance characteristics. The strain RC6b showed resistance against 350 mg Cd L^{-1}, 1000 mg Zn L^{-1} and 1200 mg Pb L^{-1}. Among the heavy metals, Pb and Zn were less toxic, whereas Cd were highly toxic to strain RC6b with the order of resistance is Pb > Zn > Cd. For more information on the behavior of the microbial strains in metal contaminated liquid medium and the capacity of the strains to survive and grow in unfavorable conditions, the growth rate of RC6b in the presence of heavy metals was also determined. The growth pattern of RC6b exhibited a variation in control compared to that of the metals used (Fig. 5).

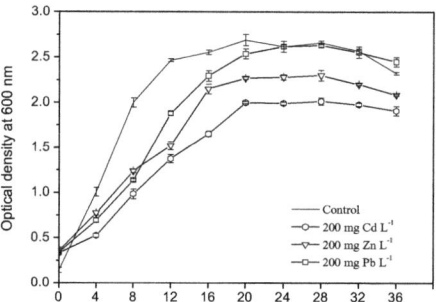

Figure 5 Growth pattern of *P. myrsinacearum* RC6b in LB medium supplemented with metals at the concentrations of 200 mg L^{-1}.

During the initial 20 h, the maximum growth was observed in control followed by that exposed to 200 mg Pb L^{-1}. Although a slight decrease in the overall growth of RC6b in the presence of metals was evident during the initial 12 h, the bacterial cells were able to return to normal growth under all conditions tested after 16 h. Similar results were also reported for other metal resistant rhizobacteria e.g., *Bacillus thuringiensis* OSM29, *Agrobacterium tumefaciens* LMG196 (Wei et al., 2009; Oves et al., 2013) indicating that the bacterial strains isolated from metal polluted soils have adapted to multiple heavy metal stress by developing various mechanisms.

3.2.3. Plant growth promoting traits of P. myrsinacearum RC6b

The plant associated bacteria isolated from metal contaminated rhizosphere soils that are known to improve the plant growth in the presence of heavy metals have various plant growth promoting traits such as production of ACC deaminase, IAA, siderophores and/or solubilization of P, which are the implicated mechanisms that contribute to the reduced metal stress and increased growth in their host plants (Ma et al., 2011a; Rajkumar et al., 2012).

Production of ACC deaminase by PGPR is one of the key traits that attenuate ethylene-mediated plant growth inhibition through metabolizing the ethylene precursor, ACC into α-KB and ammonia (Glick et al., 2007). In this study, the metal mobilizing strain RC6b was initially tested for its ability to grow on DF salts minimal medium with or without ACC. The strain RC6b grew well in DF salts minimal medium with ACC, whereas, in the absence of ACC it showed a limited growth (Fig.6).

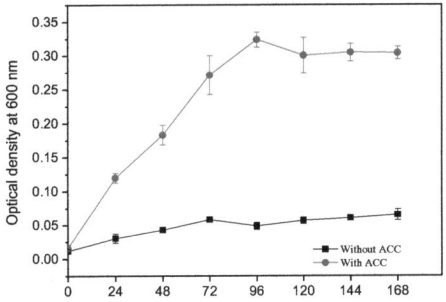

Figure 6 Growth of RC6b on DF salts minimal medium.

These observations indicate that RC6b has the potential to utilize ACC as a sole source of nitrogen through producing an enzyme ACC deaminase. Further, the ACC deaminase activity of

RC6b was analyzed by quantifying the amount of α-KB produced. The isolate produced 15.2 μmol α-KB mg^{-1} protein h^{-1}, which confirmed the enzyme activity.

Siderophore production by metal resistant PGPR is an important biological process, making iron available to plants in metal polluted soil environment (Rajkumar et al., 2012). In this study, the production of siderophores by RC6b was analyzed using CAS method. Strain RC6b exhibited positive reactions for siderophore production by forming orange-colored zone on CAS agar plates. Further the types of siderophores were also determined by the colorimetric method of Arnow (1937), using 2,3-dihydroxybenzoic acid and the Atkin et al. (1970) assay, using desferrioxamine mesylate as standards, where RC6b shown the ability to produce both catechol (654 mg L^{-1}) and hydroxamate (83.9 mg L^{-1}) type siderophores.

It has been well established that, as a common strategy to scavenge P from insoluble mineral sources, microorganisms produce and exudate various organic acids (Rajkumar et al., 2012). The phosphate solubilization potential of RC6b was studied over a time period of 120 h by monitoring pH drop and available phosphorus in the culture medium. Maximum phosphate solubilization, that is, 105 mg of P mg L^{-1} was detected after 72 h incubation along with a significant pH decrease from 4.88 to 3.73 (Fig. 7).

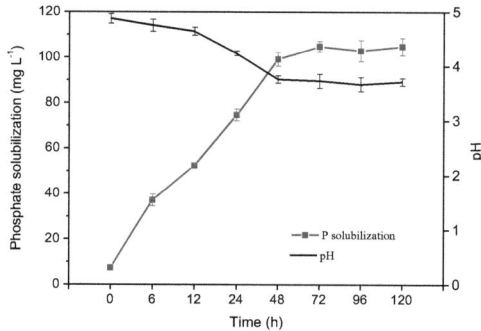

Figure 7 Phosphate solubilization by RC6b. The amount of soluble phosphates released was determined from the absorbance data using the calibration curve of KH_2PO_4 at 880 nm. Bars represent standard deviations of three replicates.

These results indicate that acidification seemed to be the main strategy followed by RC6b for solubilizing P. A recent study on the influence of phosphobacteria isolated from the rhizosphere of *Coffea arabica* L. on solubilizing insoluble hydroxyapatite/tricalcium phosphate also revealed that the solubilization of P compounds strongly depended on the release of various organic acids such as 2-ketogluconic, gluconic acids and acetic acid (Muleta et al., 2013). It has been reported that the plant associated microbes in metal polluted rhizosphere soils may mobilize insoluble phosphates very efficiently as a consequence of the production of various organic acids, which results in decrease in the metal-induced P deficiency in plants (Park et al., 2011; Muleta et al., 2013).

As it is well documented that the production of IAA by plant associated bacteria in the rhizosphere greatly contributes to the plant growth in metal polluted soils through stimulating plant root growth and the ability to take up water and nutrients, the potential of *P. myrsinacearum* RC6b to produce IAA was determined. As shown in Fig. 8a, the production of IAA by RC6b in LB medium supplemented with L-tryptophan (1 mg mL^{-1}) exhibited a maximum IAA accumulation (96.5 mg L^{-1}) at 72 h of incubation; thereafter, it was decreased and maintained constant for a period of time. This decrease was probably attributed to the release of IAA degrading enzymes such as IAA oxidase and peroxidase (Datta and Basu, 2000). Since root borne nutrients particularly L-tryptophan are considered as an important components for bacterial IAA production as well as for their growth in the rhizosphere, the strain RC6b was further tested for its ability to produce IAA in culture media supplemented with various concentrations of L-tryptophan. As shown in Fig. 8b, RC6b did not produce IAA in the absence of tryptophan in the growth medium whereas in the presence of 2 mg mL^{-1} tryptophan, it produced maximum amounts of IAA. However, a noticeable decrease in IAA production was observed at higher concentrations of L-tryptophan (3, 4 and 5 mg mL^{-1}). These results concur with the earlier observations Khamna et al. (2010) indicating that L-tryptophan at higher concentration exerts negative effects on IAA production. On the other side, some recent studies found strong linear correlation between the bacterial IAA production and L-tryptophan concentrations in the growth media (Legault et al., 2011; Patil, 2011). These contradictions require further studies to be clearly explained.

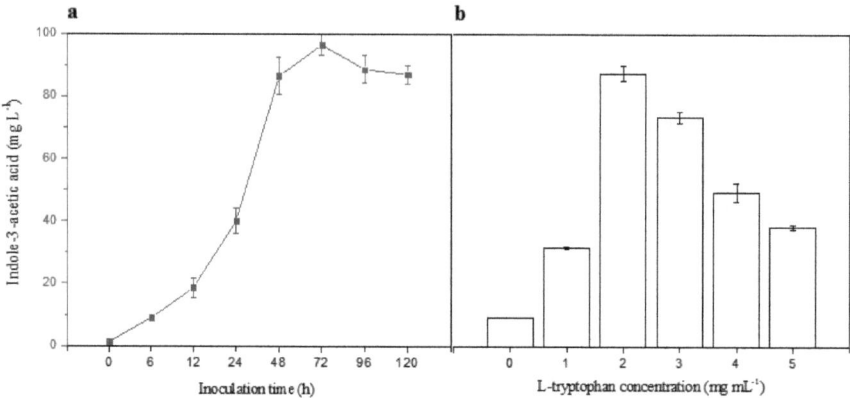

Figure 8 Effect of inoculation time (a) and L-tryptophan concentration (b) on IAA production of RC6b. Bars represent standard deviations of three replicates.

Efficient heavy metal-mobilizing abilities and the potential to grow under multi-metal stress conditions along with various plant beneficial traits are clear indications of the advantages that may offer to employ this organism as an inoculant for improving the efficiency of heavy metal phytoremediation. Similar to our findings of multiple plant growth promoting traits in metal resistant PGPR have been reported by some other workers (Sheng et al., 2008; Srivastava et al., 2013), while such findings on metal mobilizing rhizosphere isolates are less commonly explored.

3.3. Influence of P. myrsinacearum RC6b on plant growth and metal uptake

The positive effects of PGPR inoculation on heavy metal phytoremediation may be attributed to either the effect of microbial metabolites on improving plant growth or increasing plant metal uptake, or a combination of both mechanisms. In general, heavy metals in plants especially Cd even at lower concentrations, may inhibit plant growth and yield through affecting various physiological and biochemical processes (Sanita di Toppi and Gabrielli, 1999). In our study, *S.*

plumbizincicola inoculated with RC6b performed better in terms of growth in metal polluted soils (Table 2).

Table 2 Influence of *P. myrsinacearum* RC6b on the plant growth and the uptake (mg kg^{-1}) of Cd, Zn and Pb by *S. plumbizincicola*.

Treatment	Root length	Shoot length	Fresh weight	Dry weight	Cd concentration		Zn concentration		Pb concentration	
	cm	cm	g plant^{-1}	g plant^{-1}	Root	Shoot	Root	Shoot	Root	Shoot
Control	4.6±0.3	17.2±1.2	46±2	4±0	35±6	93±4	889±57	1072±38	99±11	101±11
RC6b	12.8±4.2*	21.8±1.7*	58±9*	5±1	47±5*	146±2*	1310±174*	1435±31*	15±0	5±1

Average ± standard deviation from five samples. An asterisk (*) denotes a value significantly greater than the corresponding control value according to Fisher's protected LSD test ($p < 0.05$).

The strain RC6b increased the root length, shoots length, fresh weight and dry weight by 176, 27, 27 and 22%, respectively, compared to non-inoculated plants. The increase in plant growth caused by *P. myrsinacearum* RC6b in metal contaminated soils may be attributed to its ability to produce IAA, ACC deaminase, siderophores and solubilize P (Prapagdee et al., 2013; Srivastava et al., 2013). It has been reported that PGPR (e.g., *Bacillus weihenstephanensis*, *Pseudomonas chlororaphis*, *Microbacterium lactium*, *Microbacterium* sp., *Micrococcus* sp. and *Klebsiella* sp.) isolated from metal polluted soils may help plants to produce more biomass by providing the plant with IAA that directly stimulates plant cell elongation, cell division, root initiation, and/or expression of specific genes (Prapagdee et al., 2013). Further, several plant associated bacteria were found to possess ACC deaminase suggesting their possible role in decreasing the amount of ACC as well as ethylene in the roots, thereby reducing heavy metal induced damages in plants (Glick et al., 2007). Similarly, recent studies have also indicated that under heavy metal stress conditions, inoculation with PGPR possessing the ability to produce siderophores and solubilize P increased growth of the inoculated plants primarily through enhancing the nutrient uptake in the inoculated plants (Ma et al., 2010). Our results show that RC6b can produce ACC deaminase, siderophores, IAA and solubilize P that can improve the plant growth in metal polluted soils through exhibiting individual or combined effects of plant growth promoting metabolites. Further work is under progress to elucidate the exact mechanisms that are essential to plant growth promoting potential of RC6b.

The alterations in heavy metal mobilization and its solubility in the rhizosphere soils caused by chemical and/or biological features can have dramatic effect on heavy metal uptake/accumulation in plants (Sessitsch et al., 2013). In this study thus the effects of metal mobilizing RC6b on metal accumulation in roots and shoots of *S. plumbizincicola* were studied. In general, inoculation of RC6b significantly increased the accumulation of Cd and Zn in root and shoot system (Table 2). For instance, RC6b increased Cd and Zn concentration in the shoot tissues by 57 and 34%, respectively. This corroborates the data shown in Fig. 3 for bacterial metal solubilization indicating that inoculation with metal mobilizing RC6b facilitated Cd and Zn solubilization in the rhizosphere soils and thereby enhanced their uptake by plants. Previously, Ghosh et al. (2011) also reported that the increase in arsenic bioavailability after PGPR (*Pseudomonas* sp., *Comamonas* sp. and *Stenotrophomonas* sp.) inoculation could enhance the arsenic uptake of hyperaccumulator plant *Pteris vittata* L. Prapagdee et al. (2013) also found that the inoculation of PGPR, *Micrococcus* sp. MU1 and *Klebsiella* sp. BAM1 increased Cd solubility in soils and thereby improving the phytoextraction efficiency of *Helianthus annuus* in metal polluted soils. However, in the case of Pb, RC6b inoculation decreased metal accumulation in root (85%) and shoot (95%) systems of *S. plumbizincicola* plants (Table 2) although RC6b showed Pb solubilization potential in metal solubilization experiments (Fig. 3). These effects of inoculation were also reported by Park and Bolan (2013), who found that the inoculation of plants with P-solubilizing bacteria decreased the concentration of shoot Pb in *B. juncea* in agar medium by 58.1% and in *Lolium perenne* in soil by 22.8% compared with respective non-inoculated control. This study showed that the P-solubilizing bacteria facilitate Pb immobilization via the release of P from insoluble P compounds, thus making Pb (as a carbonated fluoropyromorphite-like mineral) unavailable for plant uptake. However, Jeong et al. (2012) found that the inoculation of plants with P-solubilizing *Bacillus megaterium* increased the Cd concentration in *B. juncea* and *Abutilon theophrasti* by two folds compared with respective non-inoculated control. Taken together, present and previous research indicating that besides the bacterial metal solubilization activity, the other factors including soil nutrients level, pH, type of metals and plants, etc., greatly influence the metal solubilization in soils and thereby alter its uptake by plants (Mart ñez-Alcalá et al., 2009; Rajkumar et al., 2013a).

The efficiency of microbe-assisted phytoremediation is dependent on the survival and the competitiveness of the inoculants against native populations. Although the colonization and survival efficiency of RC6b in the rhizosphere soils has not been studied in the present study, the improved plant growth and metal accumulation (especially Cd and Zn) in plant tissues after RC6b inoculation clearly indicates its potential to tolerate, survive and express plant beneficial traits under metal stress conditions. To the best of our knowledge, this is the first work on the utilization of metal resistant PGPR RC6b as a metal mobilizer to induce phytoextraction potential of *S. plumbizincicola* in multi-metal contaminated soils.

4. Conclusions

Our work has demonstrated that metal mobilizing *P. myrsinacearum* RC6b isolated from the rhizosphere of hyperaccumulator *S. plumbizincicola*, is able to withstand high metal concentrations and can exhibit multiple plant growth beneficial properties including production of siderophores, IAA, ACC deaminase and solubilization of P. The results further suggested that activities of *P. myrsinacearum* RC6b in the rhizosphere soils can significantly improve the phytoremediation potential of plants in metal polluted soils through increasing two factors that control this parameter, i.e. plant biomass production and its metal accumulation. Further investigations on this metal mobilizing *P. myrsinacearum* RC6b for its efficiency under field conditions are in progress to promote it as bioinoculant for improving the phytoremediation in metal polluted soils.

Chapter II Biotechnological potential of hyperaccumulator-endophytic bacteria interaction with enhanced phytoextraction

Endophyte-assisted phytoremediation has recently been suggested as a successful approach for ecological restoration of metal contaminated soils, however little information is available on the influence of endophytic bacteria on the phytoextraction capacity of metal hyperaccumulating plants in multi-metal polluted soils. The aims of our study were to isolate and characterize metal-resistant and 1-aminocyclopropane-1-carboxylate (ACC) utilizing endophytic bacteria from tissues of the newly discovered Zn/Cd hyperaccumulator *Sedum plumbizincicola* and to examine if these endophytic bacterial strains could improve the efficiency of phytoextraction of multi-metal contaminated soils. Among a collection of 42 metal resistant bacterial strains isolated from the tissues of *S. plumbizincicola* grown on Pb/Zn mine tailings, five plant growth promoting endophytic bacterial strains (PGPE) were selected due to their ability to promote plant growth and to utilize ACC as the sole nitrogen source. The five isolates were identified as *Bacillus pumilus* E2S2, *Bacillus* sp. E1S2, *Bacillus* sp. E4S1, *Achromobacter* sp. E4L5 and *Stenotrophomonas* sp. E1L and subsequent testing revealed that they all exhibited traits associated with plant growth promotion, such as production of indole-3-acetic acid and siderophores and solubilization of phosphorus. These five strains showed high resistance to heavy metals (Cd, Zn and Pb) and various antibiotics. Further, inoculation of these ACC utilizing strains significantly increased the concentrations of water extractable Cd and Zn in soil. Moreover, a pot experiment was conducted to elucidate the effects of inoculating metal-resistant ACC utilizing strains on the growth of *S. plumbizincicola* and its uptake of Cd, Zn and Pb in multi-metal contaminated soils. Out of the five strains, *B. pumilus* E2S2 significantly increased root (146%) and shoot (17%) length, fresh (37%) and dry biomass (32%) of *S. plumbizincicola* as well as plant Cd uptake (43%), whereas *Bacillus* sp. E1S2 significantly enhanced the accumulation of Zn (18%) in plants compared with non-inoculated controls. The inoculated strains also showed high levels of colonization in rhizosphere and plant tissues. Results demonstrate the potential to improve phytoextraction of soils contaminated with *multiple* heavy metals by inoculating metal hyperaccumulating plants with their own selected functional endophytic bacterial strains.

1. Introduction

Hyperaccumulator plants have great potential for phytoextraction of heavy metal contaminated soils (Reeves and Baker, 2000), and several plant species (e.g., Alyssum bertolonii, Arabidopsis halleri, Solanum nigrum, Eichhornia crassipes, and Thlaspi caerulescens) have been proposed for phytoextraction of Ni, Cd, Zn and Pb (McGrath et al., 2006). The process of phytoextraction is often more time consuming than other physiochemical technologies, and it is usually limited by the low biomass and slow growth of most hyperaccumulators. To enhance the phytoextraction efficiency, functional rhizobacteria have been used in several successful studies of treatment of heavy metal contaminated soils (Ma et al., 2009, 2013; Rajkumar et al., 2013b). More recently, attention has concentrated on the bioaugmentation with metal resistant and plant growth promoting endophytic bacteria (PGPE) for enhancing phytoextraction efficiency (Ma et al., 2011b). In some cases, PGPE can promote plant growth and establishment under adverse conditions through various mechanisms including production of plant beneficial substances such as 1-aminocyclopropane-1-carboxylate (ACC) deaminase, indole-3-acetic acid (IAA), siderophores, and/or solubilization of mineral nutrients (Ma et al., 2011a). Moreover, certain PGPE could potentially reduce the phytotoxic effects, alter heavy metal availability to the plant by producing siderophores, organic acids, biosurfactants and extracellular polymeric substances (Rajkumar et al., 2009).

Sedum plumbizincicola is a newly discovered Zn/Cd hyperaccumulator from lead-zinc mining areas in Zhejiang Province, PR China (Wu et al., 2008). *S. plumbizincicola* is capable of extracting high concentrations of Zn and Cd from polluted soils, and may have great potential for phytoextraction of metal contaminated soils (Ma et al., 2013). Although the heavy metal accumulation and phytoextraction efficiency of *S. plumbizincicola* have been studied, the effects of metal-resistant and ACC utilizing endophytic bacteria on accumulation of heavy metals and nutrients in *S. plumbizincicola* grown in metal contaminated soils and PGPE-metal-hyperaccumulator interactions have not received much attention.

In this study, a novel phytoextraction system comprising hyperaccumulator plants and their endophytic bacteria was constructed for the remediation of soil contaminated with multiple metals. The objectives of our study were: i) to isolate and characterize metal-resistant and ACC utilizing endophytic bacteria from tissues of *S. plumbizincicola* and ii) to examine if the target PGPE could improve the efficiency of phytoextraction by investigating several parameters, such as plant growth, nutrients accumulation, metal mobilization and uptake by *S. plumbizincicola* in multi-metal contaminated soils. To our knowledge, this is the first report to use functional endophytic bacteria with *S. plumbizincicola* for improving phytoextraction of soils polluted with multiple metals.

2. Materials and methods

2.1. Isolation of metal resistant PGPE

The endophytic bacterial strains were isolated from fresh fine tissues (stems and leaves) of *S. plumbizincicola* grown in the vicinity of Pb/Zn mine tailings in Chunan city of Zhejiang province, China, following the method of Ma et al. (2011b). The soil where the plants were growing has the following characteristics: soil pH (1:1 w/v water) and organic matter were 7.6 and 1.36%, respectively, and total metal concentrations were 1826.7 mg Cu kg^{-1}, 991.9 mg Zn kg^{-1}; 91.3 mg Cd kg^{-1} and 14.2 g Pb kg^{-1}. Diluted tissue extracts (100 μL) were plated onto sucrose minimal salts low-phosphate (SLP) agar plates amended with 100 mg of Cd (CdCl$_2$), Zn (ZnSO$_4$), or Pb [Pb(NO$_3$)$_2$] per liter. After incubating at 37 ℃ for 5 d, morphologically distinct bacterial colonies were randomly picked, purified and restreaked on the same media until the colonies of each isolate were apparently morphologically homogeneous (Ma et al., 2009). In order to isolate PGPE, all the metal-resistant isolates were tested whether they were able to grow on DF salts minimal medium (Dworkin and Foster, 1958) with ACC as the sole nitrogen (N) source. Further, the plant growth promoting effects of the isolated endophytic bacterial strains were determined by assessing the relative elongation ratio (RER) of root and shoot, plant fresh and dry weight, and vigor index, of the model plant *Brassica napus* with or without the presence of Cd (6 μM CdCl$_2$) according to the modified assay of Ma et al. (2009).

2.2. Characteristics of PGPE

2.2.1. Genetic identification

Genomic DNA was isolated from pure bacterial colonies by using the QuickExtract™ bacterial DNA extraction kit. Primers FAM27f (5'-GAGTTTGATCMTGGCTCAG-3') (Lane, 1991) and 1492r (5'-GGYTACCTTGTTACGACTT-3') (Kane et al., 1993), a pair of highly conserved flanking sequences were used to amplify the 16S ribosomal genes under the reaction conditions described by Branco et al. (2005). PCR products were visualized on 1.5% agarose gels, and the products were excised and purified from salts and primers using the PCR purification kit (Roche Diagnostics) according to the manufacturer's instructions prior to sequencing. Partial nucleotide sequence of the amplified 16S rDNA was determined using an ABI 3130XL automatic DNA sequencer (Applied Biosystems, Perkin Elmer), and then compared to similar sequences in the GenBank using the BLASTn analysis (Altschul et al., 1997).

2.2.2. Sensitivity to metals and antibiotics

The minimal inhibitory concentrations of heavy metals (Cd, Zn and Pb) at varying concentrations (50–6000 mg L^{-1}) against isolates were determined on Luria-Bertani (LB) medium as the lowest concentration of metal ion preventing bacterial growth (Yilmaz, 2003). The antibiotic sensitivity of bacterial isolates was determined by the disc diffusion method (Rajkumar et al., 2008). A 0.1 mL of undiluted overnight grown bacterial culture [10^8 colony-forming units (CFU) mL^{-1}] was spread on LB agar plates in order to form a bacterial lawn. After incubating for 24 h at 27 °C, antibiotic discs were placed on the surface of the bacterial lawn and plates were incubated at 27 °C. The following antibiotics were used: ampicillin (10 μg), chloramphenicol (30 μg), kanamycin (30 μg), penicillin (20 μg), streptomycin (20 μg) and tetracycline (30 μg). Three replicate plates with six discs each were used for each strain-antibiotic combination. The diameter of the inhibition zone was measured after 24 h and bacterial strains were classified as resistant (R) (<10 mm), intermediate (I) (10-15 mm) and susceptible (S) (>15mm).

2.2.3. Plant beneficial features

The ACC deaminase activity of cell extracts was assayed based on the determination of α-ketobutyrate (α-KB) generated through the enzymatic cleavage of ACC according to Honma and Shimomura (1978). To measure the specific activity of the cultures, protein content was determined by the Lowry method (Lowry et al., 1951).

Bacterial IAA production ability was colorimetrically measured by mixing 4 mL of Salkowski's reagent (Gordon and Weber, 1951) with 1 mL of supernatant obtained from bacterial culture grown in LB medium with L-tryptophan (500 μg mL^{-1}) (Bric et al., 1991). The absorbance of pink color developed after 25 min incubation was read at OD$_{530 nm}$. The concentration of IAA in each sample was determined using a calibration curve of IAA ranging from 0.5 to 25 μg mL^{-1}.

The bacterial cultures were grown at 27 °C for 120 h at 175 rpm in modified Pikovskaya's medium (Sundara-Rao and Sinha, 1963) supplemented with 0.5% tricalcium phosphate (TCP). The culture supernatants were collected by centrifugation at 8000 rpm for 20 min. The soluble phosphate in the culture supernatant was quantified according to Fiske and Subbarow (1925).

The production of siderophore was qualitatively assessed in chrome azurol S (CAS) agar medium according to Schwyn and Neilands (1987). CAS agar plates were spot inoculated with bacterial cultures (10 μL of approximately 10^8 CFU mL^{-1}) and incubated at 27 °C for 5 d. The diameters of orange halo formed around colonies indicated the siderophore production level. The production of catechol and hydroxamate siderophores in culture supernatants of bacteria grown under iron-limiting conditions in casamino acids medium was also quantitatively determined following the method of Arnow (1937) and Atkin et al. (1970), respectively.

2.3. Effects of PGPE on mobility of soil metals

Soils used in the assay were obtained from metal contaminated agricultural land in Fuyang city of Zhejiang province, PR China. Five soil samples were randomly collected (0–15 cm depth) and mixed in a composite sample (Table 3).

Table 3 Physicochemical properties of soils used in the metal mobilization and pot experiments.

Soil property	pH	CEC cmol kg^{-1}	OM g kg^{-1}	Total element concentration					
				N g kg^{-1}	P g kg^{-1}	K g kg^{-1}	Cd mg kg^{-1}	Zn mg kg^{-1}	Pb mg kg^{-1}
	8.1 ± 0.3	11.4 ± 0.1	36.3 ± 1.2	1.7 ± 0.0	1.1 ± 0.1	18.6 ± 0.2	5.9 ± 0.3	736.2 ± 13.1	153.3 ± 8.2

Values are means ± standard deviations of five replicates. CEC cation exchange capacity; OM, organic matter.

The soil was air dried, crushed to pass through a 2-mm sieve and then autoclaved at 121 °C for 2 h. Pure culture of bacterial strain was incubated at 27 °C with 200 rpm shaking for 24 h. After adjusting cells to an OD_{600} of 1.5 (cuvette path length of 1 cm), 5 mL of cultures were centrifuged, then gently washed twice with 0.05 M phosphate buffer (pH 7.2) and three times with sterile deionized water, recentrifuged and finally resuspended in 5 mL sterile deionized water. One mL of washed bacterial culture was added to 1 g of sterilized soil in 50-mL sealed polypropylene centrifuge tubes. Sterile deionized water was used as the control inoculum. Tubes were weighed and kept at 27 °C with 200 rpm in the dark. The tubes were again weighed after 7 d and sterile deionized water was added to compensate for water-evaporation. Ten milliliters of sterile deionized water were added to each tube to extract the soil water-soluble metal (Ma et al., 2009). The soil suspensions were centrifuged (7000 rpm for 10 min) and filtered. A flame atomic absorption spectrophotometer (AAS) (PerkinElmer model 100, Massachusetts, USA) was used to determine the concentrations of Cd, Zn and Pb in the filtrate (Freitas et al., 2004; Ma et al., 2010). In addition, the pH of the medium was also recorded with a pH meter (P260; Beckman) equipped with glass electrode.

2.4. Extracellular enzyme production

Pectinase and cellulase activities of bacterial strain were determined using the disc plate method of Mateos et al (1992). The bacterial cultures were individually plated on nitrogen-freebase (NFB) agar medium supplemented with 0.5% pectin and carboxymethyl cellulose as the sole carbon source. The pectinase and cellulase activities were determined with bromothymol blue and congo red, respectively for measuring the size of clear zone formed around the colony.

2.5. Pot experiment

Experiments for determining the effects of the endophytic bacterial strains on plant growth and uptake of metals and nutrients were carried out in pots containing non-sterilized metal contaminated soil (Table 3).

The soil was air-dried and passed through a 2 mm sieve. The seedlings of *S. plumbizincicola* were obtained from an old Pb/Zn mine in Chunan city of Zhejiang province, PR China. Fresh shoot samples were clipped with shears approximately 5 cm long with 1 pair of leaves and 4-5 nodes, cleaned with tap water and grown in a half-strength Hoagland's nutrient solution for 7 days. Roots of precultured seedlings were surface-sterilized by sequential immersion in 70% (v/v) ethanol for 1 min, and 3% NaClO for 3 min and washed several times with sterile deionized water.

To determine the ability of endophytic bacteria to colonize and survive in plant hosts, mutants of the endophytic bacterial strains marked with antibiotic resistance were obtained after plating of the parental strains onto LB agar amended with chloramphenicol (150 mg L^{-1}) (Ma et al., 2011b). After incubation for 5 d at 28 ℃, the chloramphenicol resistant strains were selected based on similarities with the parent strains in colony morphology, metal resistance, metal solubilizing abilities and plant growth promotion and were recultured on chloramphenicol free medium to check stability of the antibiotic resistance marker.

For inoculation of the seedlings, the endophytic bacterial strains marked with antibiotic resistance were grown overnight in LB medium at 27 °C. Cells in the exponential phase were collected by centrifugation at 6000 rpm for 10 min and the pellet was washed twice with biological saline (0.85% KCl). The pellet was resuspended in biological saline and the OD_{600} was adjusted to 1.5. The roots were soaked for 2 h in the bacterial culture or sterile water (controls) and the seedlings were transplanted into plastic 1500 cm^3 pots containing 750 g of soil (six plants pot^{-1}). The plant seedlings were allowed to grow in a greenhouse at 25 ± 5 °C and a 16:8 h d/night regime. Each treatment was performed in five replicates. Plants were watered with deionized water three times per week. After 75 d, the plants were carefully removed from the pots and the root surface was cleaned several times with deionized water. To examine the introduced strains (E4L5, E1S2, E2S2, E4S1 and E1L), tissues (roots, stems and leaves) of *S. plumbizincicola* were ground up with a sterile mortar and pestle in 5 mL of sterile deionized water. Serial dilutions of plant tissue materials were spread on plates containing SLP agar with Cd (CdSO$_4$) (200 mg L^{-1}), Zn (ZnSO$_4$) (700 mg L^{-1}) and Pb [Pb(NO$_3$)$_2$] (1000 mg L^{-1}) according to metal resistance of isolates. After incubation for 7 d at 28 ℃, the reisolated, metal-resistant strains were identified for colony characteristics (morphology and color), metal-resistance (Cd, Zn and Pb) and ACC deaminase activity against the parent strains. Plant root and shoot length, fresh and dry weight were measured. The accumulation of elements (Cd, Zn, Pb and Fe) in roots and shoots was quantified by AAS after digestion with perchloric acid, sulfuric acid and distilled water at the ratio of 1:0.75:0.75 (v/v/v) according to the modified sulfuric-perchloric acid digestion method (Fenton and Fenton, 1979; Ma et al., 2010). The content of phosphorus (P) in tissues of *S. plumbizincicola* was determined by the molybdenum-ascorbic acid method (Murphy and Riley, 1962) after digestion with HNO$_3$ and HClO$_4$ (1:1 v/v).

2.6 Statistical analysis

The normality and homogeneity of variances of the data were verified using the Shapiro-Wilk and the Levene test, respectively. The *Brassica napus* growth data were analyzed using two-way analysis of variance (ANOVA) for each dependent variable (RER of root and shoot, plant fresh and dry weight and vigor index) versus the independent variables [Cd concentration (Cd) and bacterial strain (BS)]. When a significant *F*-value was obtained ($p < 0.05$), treatment means were compared using the Fisher's least significant difference (LSD) test. For the remaining data, treatment means were compared using one-way ANOVA followed by the post-hoc Fisher's LSD test ($p < 0.05$). All the statistical analyses were performed using SPSS 10.0.

3. Results and discussion

3.1. Isolation, characteristics and identification of metal resistant PGPE

A total of 42 morphologically different metal-resistant endophytic bacterial strains were isolated and repeatedly screened for their metal resistance in SLP medium supplemented with 100 mg L^{-1} of Cd, Zn or Pb. In order to screen the PGPE, all isolates were qualitatively tested for their ability to grow on DF salts minimal medium with ACC. The results showed that five isolates, designated E1S2, E4S1, E2S2, E4L5 and E1L were able to grow in DF salts minimal medium with ACC as a sole N source (data not shown). Bacterial strains possessing ACC deaminase are able to reduce ethylene production in stressed plants resulting from a decrease in its precursor ACC, thus enhancing the root elongation and the growth of plants (Belimov et al., 2009). Further, five ACC utilizing isolates were repeatedly screened for their effect on the growth of *B. napus* in Petri dishes with or without Cd (6 µM CdCl$_2$). In general, inoculation of ACC utilizing strains recorded a significant increase in RER of root and shoot, fresh and dry weight of *B. napus* (Table 4).

Table 4 Influence of 1-aminocyclopropane-1-carboxylate (ACC) utilizing endophytic bacteria on relative elongation ratio (RER) of root and shoot, fresh and dry weight, and vigor index of the model plant *Brassica napus* grown with or without Cd.

Treatment		RER of root [a]	RER of shoot [b]	Fresh weight (g)	Dry weight (g)	Vigor Index [c]
Control	Blank	-	-	2.62 ±0.04 c	0.15 ±0.01 c	875 ±74 d
	E1S2	149 ±17 b	180 ±9 b	3.33 ±0.10 ab	0.19 ±0.02 b	1332 ±94 b
	E4S1	137 ±10 bc	172 ±4 b	3.30 ±0.14 ab	0.20 ±0.02 b	1299 ±150 bc
	E2S2	174 ±6 a	205 ±16 a	3.70 ±0.55 a	0.23 ±0.02 a	1645 ±185 a
	E4L5	151 ±14 b	190 ±14 ab	3.58 ±0.10 a	0.19 ±0.02 b	1345 ±105 b
	E1L	124 ±8 c	147 ±15 c	3.10 ±0.20 b	0.17 ±0.02 bc	1079 ±184 cd
6 μM Cd	Blank	-	-	2.05 ±0.09 w	0.11 ±0.02 z	620 ±97 z
	E1S2	168 ±9 yz	186 ±9 xy	2.93 ±0.17 z	0.16 ±0.01 y	1081 ±155 xy
	E4S1	155 ±9 zw	186 ±3 xy	2.96 ±0.10 yz	0.15 ±0.03 y	1073 ±167 xy
	E2S2	183 ±8 x	210 ±23 x	3.18 ±0.04 x	0.19 ±0.02 a	1301 ±184 x
	E4L5	181 ±5 xy	201 ±14 x	3.15 ±0.09 xy	0.17 ±0.01 xy	1094 ±138 xy
	E1L	142 ±6 w	157 ±13 y	2.93 ±0.17 z	0.11 ±0.02 z	855 ±8 yz
Cd concentration (Cd)		F=26.9 ***	F=3.0 ns	F=38.5 ***	F=42.2 ***	F= 31.4 ***
Bacterial strain (BS)		F=19.2 ***	F=12.9 ***	F=23.9 ***	F=17.6 ***	F= 19.2 ***
Cd x BS		F=0.9 ns	F= 0.1 ns	F=0.8 ns	F=1.4 ns	F=0.2 ns

Values are means± standard deviations of three samples. Data of columns indexed by the same letter within each treatment (with or without Cd) are not significantly different between bacterial treatments according to Fisher's least significant difference (LSD) test ($p < 0.05$). For the F values of two-way ANOVA: ***, significant effect at the level of $p < 0.001$; ns, non-significant effect.

[a] Relative Elongation Ratio (RER) of root = Mean root length of tested plant / Mean root length of control x 100

[b] Relative Elongation Ratio (RER) of shoot = Mean shoot length of tested plant / Mean shoot length of control x 100

[c] Vigor index = germination (%) × seedling length (shoot length + root length)

However, strain E2S2 showed maximum plant growth promoting activities in both unpolluted and polluted media. For instance, in the presence of Cd, the inoculation of E2S2 enhanced RER of root and shoot by 183 and 209%, respectively; fresh and dry weight of *B. napus* by 55 and 73%, respectively; and vigor index by 109% compared to non-inoculated controls.

On the basis of morphological, physiological, biochemical characteristics (data not shown) and comparative analysis of the partial 16S rDNA sequence with already available database showed that strains E1S2 (847 bp) and E4S1 (843 bp) were close to the members of the genus *Bacillus*. Strain E2S2 (841 bp) was identified as *Bacillus pumilus* and E4L5 (855 bp) showed high similarity with *Achromobacter* sp., as did E1L (832 bp) with *Stenotrophomonas* sp. The partial 16S rRNA sequences of endophytic bacteria were deposited in the GenBank nucleotide sequence database under the following accession numbers: AY660543 (*Achromobacter* sp. E4L5), AY660542 (*Bacillus* sp. E1S2), AY660546 (*Bacillus pumilus* E2S2), AY660545 (*Bacillus* sp. E4S1), and AY660547 (*Stenotrophomonas* sp. E1L).

3.2. Plant beneficial features of ACC utilizing endophytic bacteria

Microorganisms isolated from natural contaminated environments often possess tolerance to multiple pollutants and various antibiotics, which possibly help them to adapt to such stressful environments (Pal et al., 2005). In this study, the ACC utilizing strains were found highly resistant to Cd, Zn or Pb when cultivated under increasing metal levels in the growth medium. Extremely high Cd and Zn resistance (up to the concentration of 400 mg L^{-1} and 1500 mg L^{-1}, respectively) was observed for the strain E2S2, whereas strain E1L showed *maximum resistance* to Pb (6000 mg L^{-1}) with relatively low resistance to Cd and Zn. The order of the toxicity of the metals to the strains was found to be Cd > Zn > Pb (Table 5).

Table 5 Heavy metal and antibiotic resistance of 1-aminocyclopropane-1-carboxylate (ACC) utilizing endophytic bacteria.

Strain	Minimum inhibitory concentrations			Antibiotic resistance (diameter of inhibition zone: mm)				
	Cd	Zn	Pb	Ampicillin	Tetracycline	Streptomycin	Chloramphenicol	Kanamycin
		mg L^{-1}		20 µg	30 µg	20 µg	30 µg	30 µg
E1S2	300	750	4500	1 (R)	2 (R)	0 (R)	0 (R)	2 (R)
E4S1	300	750	3500	2 (R)	1 (R)	18 (S)	0 (R)	4 (R)
E2S2	400	1500	3500	0 (R)	1 (R)	15 (I)	0 (R)	0 (R)
E4L5	250	1500	1000	6 (R)	0 (R)	3 (R)	0 (R)	2 (R)
E1L	350	750	6000	0 (R)	0 (R)	0 (R)	0 (R)	0 (R)

R, resistant (<10 mm); I, intermediate (10-15 mm); S, susceptible (>15 mm).

This high resistance to heavy metals could be attributed to the fact that the ACC utilizing strains were isolated from the tissues of the hyperaccumulator *S. plumbizincicola*, which can accumulate Cd and Zn from multi-metal contaminated mine soils (Wu et al., 2008) and therefore provide a specific niche to host endophytic bacteria. Besides, out of 5 tested antibiotics, all the ACC utilizing strains exhibited resistance against ampicillin, chloramphenicol, kanamycin and tetracycline (Table 5). It has been reported the antibiotic resistance properties of bacteria are often concomitant with heavy metal tolerance (Rajkumar et al., 2008).

The metal adaptation capability and metal biomobilization potential of endophytic bacteria can contribute to hamper the detrimental effect of metal pollution, making them promising for microbe-assisted phytoremediation (Ma et al., 2011a). IAA production was observed in all strains (Table 6).

Table 6 Plant beneficial traits of 1-aminocyclopropane-1-carboxylate (ACC) utilizing endophytic bacteria.

Strain	ACC deaminase	P solubilization	IAA synthesis	Siderophore production		
				CAS assay	Catechol	Hydroxamate
	μmol α-KB mg^{-1}h^{-1}	mg L^{-1}	mg L^{-1}	cm	mg L^{-1}	mg L^{-1}
E1S2	23.5±2.5 b	151.6±7.4 b	24.5±1.2 d	1.1±0.3 b	219.1±3.8 c	21.3±1.3 bc
E4S1	14.0±2.1 c	53.6±2.4 e	116.2±2.4 b	1.9±0.1 a	891.6±25.2 a	51.1±0.4 a
E2S2	32.6±4.8 a	64.1±4.3 d	128.7±4.2 a	1.5±0.3 ab	420.8±6.4 b	26.3±1.4 b
E4L5	14.8±2.6 c	166.7±3.7 a	22.8±0.5 d	1.0±0.2 b	216.6±1.5 c	20.1±0.3 c
E1L	11.5±1.0 c	126.9±9.8 c	45.3±2.6 c	1.1±0.2 b	215.8±7.5 c	20.7±0.2 bc

Values are means± standard deviations of three samples. Data of columns indexed by the same letter are not significantly different between bacterial treatments according to Fisher's least significant difference (LSD) test ($p < 0.05$). α-KB, α-ketobutyrate; P, phosphorus; IAA , indole-3-acetic acid; CAS, chrome azurol S.

Strain E2S2 produced the highest amount, 128.7 mg L^{-1} of IAA, followed by strain E4S1, which produced 116.2 mg L^{-1} of IAA. E1S2, E4L5 and E1L produced similar amounts of IAA, i.e., 24.5, 22.8, and 45.3 mg L^{-1}, respectively. It has been well documented that the bacterial IAA contribute much to the plant growth and development by regulating cell division and differentiation (Berleth and Sachs, 2001). This is largely supported by our findings. Increases in RER of root and shoot as well as biomass of *B. napus* (Table 4) are clearly correlated to the maximum biosynthesis of IAA by *B. pumilus* E2S2 (Table 6). Moreover, all five strains exhibited P-solubilizing ability by utilizing the insoluble TCP *as the sole source of* P in modified Pikovskayas medium (Table 6). The decolorization of bromophenol blue is due to a decrease in pH caused by the excretion of organic or inorganic acids, which is considered to be responsible for P solubilization (Rashid et al., 2004). The strain E4L5 recorded the maximum solubilization of P followed by E1S2. It has been demonstrated that the elevated levels of metals in soil interfere with uptake of nutrients such as P and can lead to plant growth retardation. This deficiency can be compensated by the P-solubilizing ability of PGPE strains (He et al., 2013). Besides, most of the commonly known ACC deaminase and siderophore-producing bacteria can prevent the ethylene-induced inhibition of root elongation through hydrolytic cleavage of ACC and suppressing the ACC synthase activity (Belimov et al., 2009). In our study, strain E2S2 showed the highest ACC deaminase activity (32.6 μmol α-KB mg^{-1} h^{-1} protein) (Table 6), by which it exerts the optimum of plant growth promoting traits under both Cd stress and unstressed condition (Table 4). Further, all five strains also displayed a positive siderophore activity indicated by the formation of orange colored zone on CAS agar plates. Strain E4S1 showed the ability to produce the highest levels of catechol and hydroxamate siderophores by 891.6 and 51.1 mg L^{-1}, respectively (Table 6).

3.3. Effect of PGPE on heavy metal mobility

In general, significant increase ($p < 0.05$) in water-soluble Cd and Zn from the metal contaminated soils by the five strains was observed compared to the control treatments. This was associated with a pH decrease (Fig. 9a and 9b).

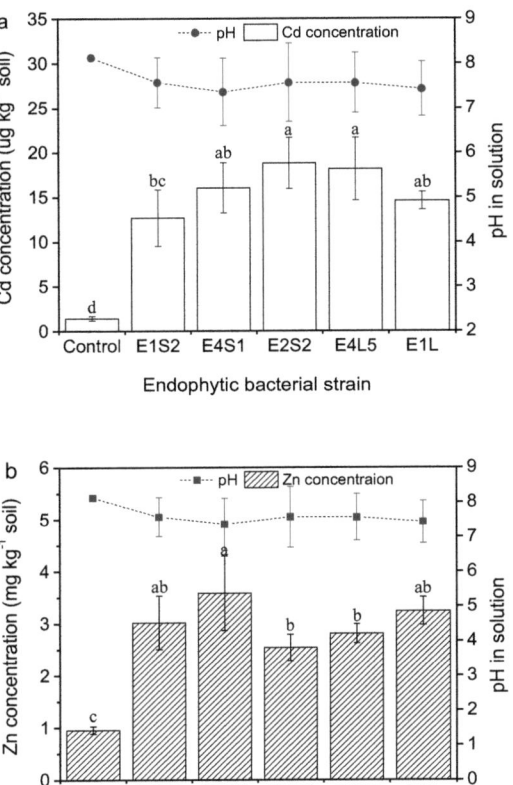

Figure 9 Effect of inoculation with endophytic bacteria on the mobilization of Cd (a), Zn (b) and Pb (c) in soil and on the pH of the metal extract solution. Each value of concentration and pH is the mean of three replicates. Error bars represent standard deviation. Data of columns indexed by the same letter are not significantly different according to Fisher's least significant difference (LSD) test ($p < 0.05$).

The maximum concentrations were typically observed for Cd (Fig. 9a). Inoculation of E2S2 significantly increased the concentrations of water extractable Cd, Zn and Pb in soil by 13.3-, 4.6- and 1.2-folds, respectively relative to the control treatment. The increased soil metal mobility can probably be attributed to acidification and siderophore production by strain E2S2, which facilitated metal solubility in multi-metal contaminated soils. Siderophore production has been proved to be stimulated by the presence of heavy metals (Gaonkar and Bhosle, 2013), and most of them, either catechols or hydroxamates, show strong affinity to divalent metal ions (Ma et al., 2009). Consequently they can possibly affect heavy metal bioavailability in soils as well. For instance, it was reported that metal-resistant *Psychrobacter* sp. SRA1 and *Bacillus cereus* SRA10 increased the amount of Ni extracted from soil, which may be due to their ability to produce maximum siderophores both in catechol and hydroxamate types (Ma et al., 2009). Besides, He et al. (2013) reported that the addition of the metal-resistant endophyte Rahnella sp. JN6 to metal contaminated soils significantly increased the water soluble Cd, Pb and Zn by 4.73, 6.21 and 7.19 folds compared with the control, respectively.

3.4. Effect of PGPE on plant growth and heavy metal uptake

As indicated in Table 3, the concentrations of Cd and Zn were very high in the metal contaminated soil used in metal mobilization and pot experiments. The concentration of total Cd and Zn were up to 5.9 mg kg^{-1} and 736 mg kg^{-1}, respectively, which are much higher than the limits of the national second-class standard (0.3 mg Cd kg^{-1} and 250 mg Zn kg^{-1}) in the People's Republic of China environmental quality standard for soils (NSPRC, 1995). Inoculated and non-inoculated control plants were subjected to the metal contaminated soil (Table 3) for 75 d and showed significant growth differences (Table 7).

Table 7 Effect of 1-aminocyclopropane-1-carboxylate (ACC) utilizing strains on the biomass of *Sedum plumbizincicola*.

Treatment	Root length	Shoot length	Fresh weight	Dry weight
	cm	cm	g plant^{-1}	g plant^{-1}
Control	4.6 ±0.3 c	17.2 ±1.2 c	44.2 ±1.1 c	4.4 ±0.1 c
E1S2	8.7 ±1.3 b	20.6 ±0.9 b	58.4 ±4.4 ab	5.5 ±0.6 a
E4S1	9.1 ±1.0 b	20.2 ±0.4 bc	54.9 ±2.8 b	4.9 ±0.2 b
E2S2	11.3 ±1.9 a	20.1 ±2.2 bc	60.4 ±2.2 a	5.8 ±0.5 a
E4L5	8.7 ±1.7 b	23.0 ±3.3 a	59.7 ±4.5 ab	5.6 ±0.5 a
E1L	9.5 ±1.1 b	19.7 ±1.0 bc	59.0 ±3.4 ab	5.5 ±0.3 a

Values are means± standard deviations of five samples. Data of columns indexed by the same letter are not significantly different between bacterial treatments according to Fisher's least significant difference (LSD) test ($p < 0.05$).

S. plumbizincicola inoculated with ACC utilizing strains exhibited significant growth increases. The highest plant growth promoting effect was found for strain E2S2, which enhanced root and shoot length, plant fresh and dry weight by 146, 17, 37 and 32%, respectively. The increase in plant growth caused by the PGPE strains under non-sterile conditions may be attributed to their ability to produce ACC deaminase, IAA, siderophores and P solubilization activity (Ma et al., 2011b; He et al., 2013). The results obtained here clearly indicate that inoculation with strain E2S2 was highly efficient at protecting *S. plumbizincicola* from growth inhibition caused by toxic metal concentrations in soil, although *S. plumbizincicola* has great potential to tolerate high metal concentrations and grow in metal contaminated soil. This result is in agreement with a previous report describing increased biomass production of Brassica napus inoculated with ACC utilizing strains (e.g., *Ralstonia* sp. J1-22-2, *Pantoea agglomerans* Jp3-3, and *Pseudomonas thivervalensis* Y1-3-9) grown in Cu-contaminated substrates (Zhang et al., 2011).

Since the process of phytoextraction by different plants depends on both plant biomass yield and metal uptake/accumulation capacity (Ma et al., 2011a), we also assessed whether inoculation of *S. plumbizincicola* plants with beneficial endophytic bacteria affected the uptake of metals (Cd, Zn and Pb) (Fig. 10).

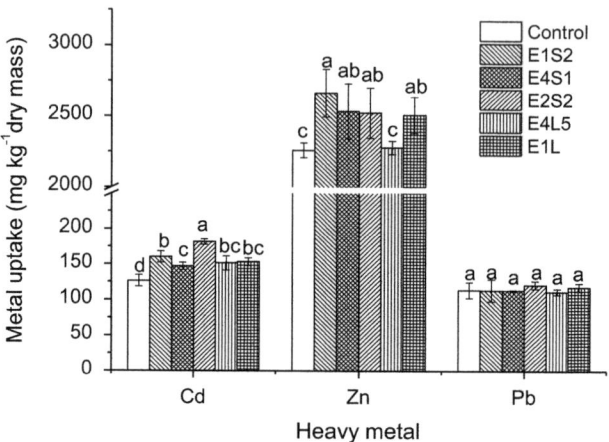

Figure 10 Influence of endophytic bacteria on the metal concentration in *Sedum plumbizincicola* grown in metal contaminated soil. Each value is the mean of triplicates. Error bars represent standard deviation. Data of columns indexed by the same letter are not significantly different according to Fisher's least significant difference (LSD) test ($p < 0.05$).

In general, the total uptake of Cd and Zn by *S. plumbizincicola* plants was significantly increased by the ACC utilizing endophytic bacterial strain E1S2, E4S1, E2S2 and E1L, compared to non-inoculated controls (Fig. 10). The inoculation of the best Cd mobilizer E2S2 significantly increased the accumulation of Cd in *S. plumbizincicola* by 43% compared with non-inoculated controls. Similarly, the maximum increase (18%) in accumulation of Zn in plants was observed when plants were inoculated with strain E1S2. The increased metal uptake induced by endophytic bacteria can be attributed to their capability of solubilizing metals in the rhizosphere of the inoculated plants through bacterial metabolic activities or their interactions with host plants (He et al., 2013). The acidification (lowering of soil pH) of surroundings resulting from P solubilization may play a key role on the solubility and accumulation of heavy metals (Rajkumar et al., 2009; Ma et al., 2011a). Several observations have evidenced that metal-resistant endophytic bacteria can alter the metal uptake capacity of hyperaccumulator plants (Visioli et al., 2014; Zhu et al.,

2014). However, bacterial inoculation did not significantly influence Pb accumulation in plants ($p > 0.05$). This observation indicates that other mechanisms, e.g. a direct dilution of metal concentrations by increased plant biomass rather than bacterial metal mobilization were involved. The application of the metal-resistant ACC utilizing endophytic bacteria not only effectively promoted the plant growth, but also increased the bioavailability of metals in the rhizosphere and consequently enhanced the uptake of Cd and Zn by *S. plumbizincicola*. Although PGPE have been used as bioinoculants in phytoremediation scenarios, most of the studies on their effects on plant growth and metal uptake have only been performed in sterile metal polluted soils or growth media (Ma et al., 2011a; He et al., 2013; Visioli et al., 2014; Zhu et al., 2014).

Soils contaminated with heavy metals often present other plants stresses, such as nutrient (e.g. Fe, P, Mg and Ca) deficiency (Ma et al., 2010; Wan et al., 2012). It has been well documented that PGPE help plants to acquire Fe and P though the activities of siderophore production and P solubilization, which can be important contributions to the beneficial effect of bacteria on plant establishment and subsequent growth in metal contaminated sites (Ma et al., 2011a,b). In our study, inoculation of all of the ACC utilizing strains did not significantly change the Fe contents in tissues of *S. plumbizincicola*, while significant increases in P contents in plant shoots were detected after bacterial inoculation, when compared with control treatments ($p < 0.05$) (Table 8).

Table 8 Influence of 1-aminocyclopropane-1-carboxylate (ACC) utilizing strains on Fe and P content in *Sedum plumbizincicola*.

Treatment	Fe content (mg kg $^{-1}$ dry mass)		P content (mg kg $^{-1}$ dry mass)	
	Shoots	Roots	Shoots	Roots
Control	500.3±82.3 a	6045.0±890.9 ab	77.8±4.3 c	129.4±20.5 ab
E1S2	472.7±20.2 a	5698.7±342.0 bc	102.8±3.6 a	177.2±54.5 a
E4S1	566.5±50.4 a	4925.8±407.7 c	102.6±7.7 a	57.9±2.4 b
E2S2	522.3±87.7 a	6129.2±451.4 ab	89.9±12.0 b	202.8±28.2 a
E4L5	548.3±65.9 a	6890.9±65.9 a	101.4±5.1 a	145.5±8.4 a
E1L	515.1±67.5 a	5878.0±610.1 b	103.8±8.4 a	169.4±33.9 a

Values are means± standard deviations of five samples. Data of columns indexed by the same letter are not significantly different between bacterial treatments according to Fisher's least significant difference (LSD) test ($p < 0.05$).

Dissimilarly, several previous studies reported increase in uptake of nutrients (Fe or P) by plants as a result of inoculations of siderophore producing or P-solubilizing bacteria (Biari et al., 2008; Ma et al., 2010).

Although the inoculated ACC utilizing endophytic strains possess several traits which facilitate the plant growth and metal uptake, as successful plant beneficial inoculants, bacteria must be able to rapidly colonize the root system and survival in metal stress environment during the growing season (Rajkumar et al., 2009). In the present study, the five metal-resistant ACC utilizing endophytic bacteria tested were able to colonize the rhizosphere or plant tissue interiors of *S. plumbizincicola*. The numbers of the bacterial colonies in the rhizosphere and tissues of *S. plumbizincicola* plants were 10^4–10^5 CFU g^{-1} of fresh rhizospheric soil, 10^3 CFU g^{-1} of fresh root and stems, and 10^2 CFU g^{-1} of fresh leaves (Table 9).

Table 9 Colonization of 1-aminocyclopropane-1-carboxylate (ACC) utilizing strains in the rhizosphere soil and tissue interior of *Sedum plumbizincicola* after inoculation with seedlings grown in metal contaminated soil.

Treatment	Rhizosphere	Root interior	Stem interior	Leaf interior
E1S2	102.3 ±6.8 bc	3.2 ±0.4 c	1.0 ±0.4 b	0.4 ±0.1 b
E4S1	89.7 ±8.5 c	2.9 ±0.2 cd	1.3 ±0.3 ab	0.4 ±0.0 b
E2S2	143.3 ±13.0 a	4.4 ±0.3 a	1.7 ±0.2 a	0.6 ±0.0 a
E4L5	112.0 ±12.1 b	3.8 ±0.1 b	1.3 ±0.4 ab	0.4 ±0.1 b
E1L	63.7 ±7.4 d	2.7 ±0.3 d	1.1 ±0.2 b	0.3 ±0.1 c

Values are means ± standard deviations of five samples. Data of columns indexed by the same letter are not significantly different between bacterial treatments according to Fisher's least significant difference (LSD) test ($p < 0.05$). Values in 10^3 CFU g^{-1} of fresh soil or plant tissue.

Strain E2S2 showed higher level of colonization in tissue (leaf, stem and root) interior and rhizosphere of *S. plumbizincicola*, compared to the other endophytic strain inoculation treatments. In addition, the production of plant cell wall degrading enzymes by endophytes such as cellulase and pectinase was determined, since these extracellular enzymes may confer an advantage of intercellular ingress and facilitate vertical spreading of endophytes into the host plants, which has been considered as an important mechanism for endophytic colonization (Verma et al., 2001). Among the five strains tested, E2S2 exerted a positive cellulase activity by the development of yellow-colored zone on NFB plates. However, except for strain E1L, the other four strains exhibited positive pectinase activity. Ma et al. (2011b) recently reported that the production of bacterial cellulase and pectinase enzymes facilitated colonization of endophyte *Pseudomonas* sp. A3R3 in tissues of both host plant *Alyssum serpyllifolium* and non-host plant *Brasscia juncea* and suggested that production of extracellular enzymes by endophytes plays an important role in plant–microbe interactions and intercellular colonization of plant tissues.

4. Conclusions

Our research demonstrated that metal-resistant ACC utilizing endophytic bacteria isolated from the tissues of *S. plumbizincicola* plants grown in a metal contaminated mine soil, are able to adapt and survive at high concentration of metals and possess various plant beneficial features including production of plant growth promoting substances such as IAA, siderophore and ACC deaminase or/and solubilization of P, as well as production of extracellular enzymes such as pectinase and cellulase. Further, our results showed that inoculation of ACC utilizing strains, especially *B. pumilus* E2S2 significantly increased mobilization of Cd and Zn in soil and subsequent uptake of these metals in the tissues of *S. plumbizincicola*, together with concurrent stimulation of plant growth. Further investigations on the efficiency of this ACC utilizing endophytic bacterial strain as a bioinoculant for improving the phytoextraction of multi-metal polluted soils under field conditions are in progress.

Chapter III Role of plant growth-promoting rhizobacterium *Bacillus* sp. SC2b in rhizoremediation

A plant growth-promoting bacterial (PGPB) strain SC2b was isolated from the rhizosphere of *Sedum plumbizincicola* grown in lead (Pb)/zinc (Zn) mine soils and characterized as *Bacillus* sp. based on (1) morphological and biochemical characteristics and (2) partial 16S ribosomal DNA sequencing analysis. Strain SC2b exhibited high-levels of resistance to cadmium (Cd) (300 mg/L), Zn (730 mg/L) and Pb (1400 mg/L). This strain also showed various plant growth-promoting (PGP) features such as utilization of 1-aminocyclopropane-1-carboxylate, solubilization of phosphate, production of indole-3-acetic acid and siderophore. The strain mobilized high concentration of heavy metals from soils and exhibited different biosorption capacity towards the tested metal ions. Strain SC2b was further assessed for PGP activity by phytagar assay with a model plant *Brassica napus*. Inoculation of SC2b increased the biomass and vigor index of *B. napus*. Considering such potential, a pot experiment was conducted to assess the effects of inoculating the metal-resistant PGPB SC2b on growth and uptake of Cd, Zn and Pb by *S. plumbizincicola* in metal-contaminated agricultural soils. Inoculation with SC2b elevated the shoot and root biomass and leaf chlorophyll content of *S. plumbizincicola*. Similarly, plants inoculated with SC2b demonstrated markedly higher Cd and Zn accumulation in the root and shoot system indicating that SC2b enhanced Cd and Zn uptake by *S. plumbizincicola* through metal mobilization or plant-microbial mediated changes in chemical or biological soil properties. Data demonstrated that the PGPB *Bacillus* sp. SC2b might serve as a future biofertilizer and an effective metal mobilizing bioinoculant for rhizoremediation of metal polluted soils.

1.   Introduction

Rapid industrialization, overuse of agrochemicals and minimal environmental protection over the past three decades resulted in significant environmental problems worldwide (Li et al., 2014). In particular, heavy metal pollution of soils due to intensified exploitation of mineral resources and emission in smelting process has become a serious concern in many developing countries. Approximately 2×10^7 ha of arable land in China has been contaminated with heavy metals such as arsenic (As), cadmium (Cd), chromium (Cr), lead (Pb) and zinc (Zn) to various degrees due to local mining and refinery activities (Li, 2005). Accumulation of heavy metals in soils and sediments not only affects soil fertility, crop yield and quality of agriculture products, but also negatively impact human and animal health by entering the food chain (Coelho et al., 2012; Dahshan et al., 2013: Figueiredo et al., 2014a; 2014b).

Although traditional physicochemical technologies used for remediation of metal polluted soils have been well developed, these approaches have many problems, including high cost, adverse effect on soil structure, fertility and biological activity (Holden, 1989). In recent years, plant based techniques such as phytoremediation, using metal tolerant and/or hyperaccumulating plants have been proposed as an environmentally friendly and low cost technology for metal stabilization and extraction from polluted soils. However, the major issues hampering the efficiency of these processes are that high biomass producing plants usually exhibit low metal tolerance and uptake potential, whereas natural hyperaccumulators generally produce low biomass and display slow growth rate (Glick, 2010).

Rhizoremediation is one of the phytoremediation methods, which depends upon interactions between plants, microbes and pollutants, where exudates released by plant roots help stimulate survival and activity of bacteria, which improve soil chemical and physical properties, enhance nutrient acquisition, metal detoxification and alleviation of biotic/abiotic stress in plants (Kuiper et al. 2004; Kamaludeen and Ramasamy, 2008). In general, the success of the rhizoremediation process in metal polluted soils depends upon the type and bioavailability of pollutants, plants, diversity and activity of microbes, and environmental conditions in the rhizosphere (Wenzel, 2009). In this context, metal mobilizing bacteria may be potential candidates for enhancing rhizoremediation, since these microbes produce various metal mobilizing metabolites including organic acids, siderophores, or biosurfactants, and thus increase concentrations of bioavailable heavy metals in the rhizosphere, and enabling them to be available for plant uptake (Ma et al., 2011a). Similarly, the effectiveness of rhizoremediation may also be improved by use of plant growth-promoting bacteria (PGPB) as beneficial inoculants, which improve plant nutrient acquisition, reduce metal toxicity, and improve plant health and biomass production under adverse environmental conditions. PGPB exert their beneficial effects on host plants though various mechanisms, such as (1) utilization of 1-aminocyclopropane-1-carboxylate (ACC, the ethylene precursor) produced by plants under stress condition as sole energy source, (2) synthesis of phytohormones including auxins, gibberellins and cytokinins that enhance plant growth and

44

development, and (3) production of different types of siderophores (catechol and hydroxamate) that solubilize and sequester available iron (Fe) from the soil, and mitigate nutrient deficiency by fixation of nitrogen and solubilization of phosphorus and potassium (Ma et al., 2011a). Ahemad and Kibret (2014) demonstrated that plants inoculated with metal-resistant PGPB are usually more tolerant to certain metals than non-inoculated plants. Considering the potential of beneficial bacteria to promote plant growth and heavy metal accumulation in plants, it may be envisaged that increasing the population density of metal-resistant beneficial bacteria in the rhizosphere might alleviate metal stress in plants and elevate bioavailable metal concentration for plant uptake, thus enhancing overall rhizoremediation process in polluted soils.

Sedum plumbizincicola (Crassulaceae) is a newly discovered Zn/Cd hyperaccumulator with high potential for rhizoremediation of metal polluted soils (Jiang et al., 2010; Wu et al., 2013). Although several recent investigations addressed rhizosphere and endophytic microbial population of plants and their role on rhizoremediation in metal polluted soils (Ma et al., 2011b; Rajkumar et al. 2013), the beneficial interactions between *S. plumbizincicola* and PGPB, and their possible role on the rhizoremediation of multi-metal polluted agricultural soils still remains poorly understood. The objectives of this study were to: i) isolate and characterize a beneficial bacterial strain possessing the ability to mobilize metals in soils, biosorb heavy metals in their cells and produce various plant growth-promoting (PGP) metabolites, such as indole-3-acetic acid (IAA), ACC deaminase and siderophores, and solubilizing phosphate; and ii) elucidate the effects of inoculation of isolated metal mobilizing PGPB strain on plant growth and rhizoremediation capacity of *S. plumbizincicola* in multi-metal polluted agricultural soils.

2.   Materials and methods

2.1.   Isolation, identification and phylogenetic analysis

Bacterial strains were isolated from the rhizosphere of *Sedum plumbizincicola* X.H. Guo et S.B. Zhou ex L.H. Wu grown on a Pb/Zn mine area in Zhejiang province, China. To isolate the metal-resistant bacteria, soil samples were serially diluted in sterile deionized water and plated on Luria-Bertani (LB) agar supplemented with 100 mg/L heavy metals $CdSO_4 \cdot H_2O$, $PbCl_2$ and $ZnSO_4 \cdot 7H_2O$ alone and in combinations. The plates were incubated at 28 °C for 48 hr. To test the level of resistance to metals, isolated bacterial strains were grown in LB agar medium supplemented with different concentration of Cd, Pb and Zn. The bacterial strain resisting highest levels of metals was selected and identified based on morphological, physiological, biochemical characteristics and 16S rRNA gene sequencing method. The physiological characteristics such as temperature, pH and salt tolerance of bacterial strain were examined using standard procedures (Mishra et al., 2009). The genomic DNA was isolated using the QuickExtract bacterial DNA extraction kit and the 16S rRNA gene was amplified by polymerase chain reaction (PCR) using the conserved eubacterial primers FAM27f (5'-GAGTTTGATCMTGGCTCAG-3') and 1492r (5'-GGYTACCTTGTTACGACTT-3'). Each amplification mixture (5 µL) was analyzed by agarose gel (1%, w/v) electrophoresis in TAE buffer (0.04 M TRIS acetate, 0.001 MEDTA) containing 1 µg/mL (w/v) ethidium bromide. Partial sequence of the PCR-amplified 16S rDNA was performed using an ABI 3130XL automatic DNA sequencer (Applied Biosystems, Perkin Elmer), and then compared with similar sequences in the NCBI GenBank database using the BLASTn program. Sequences of the 8 most closely related microorganisms and that of *Escherichia coli* were used as comparison to construct a phylogenetic tree. Bacteria used in the construction of the phylogenetic tree with their GenBank accession numbers included *Bacillus megaterium* MB1-42 (KJ843149.1), *B. horikoshii* RB10 (GU232770.2), *B. flexus* NY-1 (EU869200.1), *Bacillaceae bacterium* MSB06 (FJ189761.1), *B. aryabhattai* NN31 (KJ542774.1), *B. weihenstephanensis* MC67 (DQ345791.1), *B. cereus* PR15 (JQ435675.1), *B. thuringiensis* BAB-Bt2 (AM293345.1) and *E. coli* (J01859). The phylogenetic tree was constructed using the neighbor joining method using MEGA version 6.06.

2.2.   Biochemical characterization

Antibiotic including ampicillin, chloramphenicol, kanamycin, streptomycin, and tetracycline resistance of bacterial strain was analyzed by disc diffusion method (Rajkumar et al., 2008). The bacterial IAA production *in vivo* was determined by colorimetric measurement with the optical density (OD) at 530 nm using Salkowski's reagent as described by Patten and Glick (2002). In order to optimize the production of IAA, the bacterial strain was grown in LB broth with different concentration of L-tryptophan (0, 1, 2, 3, 4 or 5 mg/mL) or with L-tryptophan (0.5 mg/mL) at various pH ranging from 4 to 10. The ACC deaminase activity was measured in bacterial extracts using a modified protocol based on Honma and Shimomura (1978). Cells were collected and washed with 0.1 mol/L TRIS-HCl (pH 8.5) and resuspended in 1.5 mL lysate buffer. Cells were lysed on ice by sonication and centrifuged at 4500 g at 4 °C for 15 min. The protein content of the extracts was determined by the method of Bradford (1976). Quantitative estimation of phosphate solubilization was performed by inoculating 1 mL of bacterial suspension with an OD_{600} of 1.5 in 50 mL modified Pikovskaya's medium (Sundara-Rao and Sinha, 1963) with 0.5% tricalcium phosphate for 5 days. Phosphate content in supernatant was spectrophotometrically estimated by the modified Fiske and Subbarow method (1925) and quantified by comparison with a standard curve derived from known phosphorus concentrations. Siderophore production by the isolate was qualitatively detected by the Chrome Azurol S assay (Schwyn and Neilands, 1987). The bacterial isolate was further quantitatively assayed for the production of catechol and hydroxamate siderophores by inoculating the isolate in casamino acids medium under Fe-limiting conditions based upon method of Arnow (1937) and Atkin et al. (1970), respectively. Bacterial hydrogen cyanide (HCN) production was assayed by the qualitative method of Bakker and Schippers (1987). The bacterial isolate was grown on LB agar supplemented with glycine (4.4 g/L) in Petri plates with lids fitted with Whatman no. 1 filter paper previously soaked in 0.5% picric acid and 2% sodium carbonate. The changes in the color of the filter paper from orange to red indicated HCN production.

2.3. Biosorption of metals

The bacterial isolate was grown in LB medium with shaking at 27 °C until the OD_{600} reached 1.5. The cells were centrifuged for 10 min at 4500 g and obtained pellet washed three times with deionized sterile water. The harvested biomass was re-suspended in Eppendorf microtubes containing 1.5 mL of metal solution at the concentration of 50, 100 or 150 mg/L (Cd, Pb or Zn). After incubation at room temperature for 8 hr, cells were harvested by centrifugation at 4500 g. The residual metal ions in the supernatant were determined by atomic absorption spectrophotometry (AAS) (Varian SpectrAA 220FS, 220Z; Varian, Palo Alto, CA, USA). Total biosorbed metal values were calculated by taking differences between metal contents in supernatant at time zero and at time of sampling.

2.4. Biomobilization of metals

Soil samples were collected from a multi-metal contaminated agricultural site in Fuyang city, Zhejiang province, China and sterilized at 100 °C by steaming for 1 hr on three consecutive days.

The physicochemical properties of the soil were: pH (1:1 w/v water) 8.1; organic matter 36.3 g/kg; cation exchange capacity 11.4 cmol/kg; total metal concentration 5.9 mg/kg Cd, 1236 mg/kg Zn, 153 mg/kg Pb; total nutrient concentration 1.7 g/kg N, 1.1 g/kg P, 18.6 g/kg K. A pure culture of the bacterial strain grown in LB broth for 18 hr was centrifuged, washed, and re-centrifuged at 4500 g. One mL aliquot of the washed bacterial cells with *OD adjusted* to 1.5 was added to 1 g soil. Sterile deionized water was used as control. Samples were weighed and kept at 27 °C on an orbital shaker at 120 g in the dark. Sterile deionized water was added to compensate for daily evaporation. After 10 days, 10 mL sterile deionized water were added to extract metals from the soil. The soil suspensions were centrifuged at 6500 g for 10 min and filtered. Concentrations of Cd, Pb and Zn in the filtrate were determined by AAS.

2.5. Phytagar assay

A phytagar assay was used to assess PGP potential of isolated bacterial strain under metal stress condition. *Brassica napus* was used as a model plant because of rapid growth and high biomass production in short duration (Dell'Amico et al., 2008; Zhang et al., 2011). Surface-sterilized seeds of *B. napus* were inoculated with the PGPB and allowed to grow in phytagar medium containing 5 mg/L metal (Cd, Pb and Zn) according to the modified assay detailed by Ma et al. (2011b). Germination rate, shoot and root length, plant dry weight and the elongation rate of shoot and root and vigor index were calculated.

2.6. Pot experiment

A pot experiment was performed using the above described multi-metal contaminated agricultural soil. Soil was air-dried and passed through a 2 mm nylon sieve. Seedlings of *S. plumbizincicola* were obtained from a Pb/Zn mine area in Chunan city, Zhejiang province, China. Shoot samples (approximately 5 cm long) were cleaned with deionized water and grown in a half-strength Hoagland's nutrient solution for one week. The roots of the pre-cultured seedlings were disinfected by immersion in 70% (v/v) ethanol for 1 min and in 3% NaClO for 3 min and washed several times with sterile deionized water. For inoculation of seedlings, mutants of PGPB marked with antibiotic resistance were obtained after plating of parental strain onto LB agar containing chloramphenicol (150 mg/L). Cells grown in LB medium for 18 hr at 27 °C were collected by centrifugation at 4500 g for 10 min and the pellet was washed twice with biological saline (0.85% KCl). The pellet was re-suspended in saline and the OD_{600} was adjusted to 1.5. The surface-sterilized roots were soaked for 2 hr in the bacterial culture or sterile water (controls) and the seedlings were transplanted into plastic pots containing 750 g metal-contaminated soil (6 plants/pot). The seedlings were allowed to grow in a greenhouse at 25 ℃ and a 16:8 hr light/dark cycle. Each treatment was performed in 5 replicates. After 75 days, plants were removed from pots and root surface cleaned several times with deionized water. Plant root and shoot length, fresh and dry weight were measured. The concentrations of chlorophyll (Chl) a, Chl b and total Chl were determined in the leaves according to the method of Inskeep and Blooom (1985). The concentrations of Cd, Pb and Zn in the root and shoot system were determined by AAS (Ma et al., 2011b). To examine the colonization of introduced bacterial strain, rhizosphere samples (1 g) of *S. plumbizincicola* were collected and suspended in 10 mL sterile deionized water. Serial dilutions were prepared and spread on LB agar plates containing 150 mg/L chloramphenicol. After incubation for 7 days at 28 ℃, chloramphenicol resistant colonies were counted and expressed as colony forming units (CFU) and compared for colony characteristics (morphology and color), metal resistance and PGP traits against the parent strains.

2.7. *Statistical analysis*

Data were analyzed using Student's *t*-test ($p < 0.05$). Correlations were determined by linear regression analysis by the least square method ($p < 0.05$). All the analyses were performed using SPSS 10.0.

3. Results

3.1. *Characterization of the bacterial strain*

A total of 45 metal-resistant bacterial strains were initially isolated from the rhizosphere of *S. plumbizincicola* and strain SC2b was specifically selected due to its high resistance to Cd (300 mg/L), Pb (1400 mg/L) and Zn (750 mg/L). Strain SC2b was a Gram positive, endospore-forming, rod-shaped bacterium. The isolate was able to grow at a wide temperature range from 4 to 37 °C; however maximal growth occurred at 26 °C. The strain also showed ability to grow over a wide range of pH (5-9) and tolerated a NaCl concentration of up to 5%. Although it was susceptible to tetracycline (30 µg/mL), kanamycin (30 µg/mL), and streptomycin (20 µg/mL), this strain exhibited resistance to high concentrations of ampicillin (500 µg/mL), penicillin (300 µg/mL) and chloramphenicol (150 µg/mL). Based on the morphological, physiobiochemical characteristics (data not shown) and 16S rRNA gene sequencing analysis, evidence indicated that strain SC2b was related to the *Bacillus* genus. The highest sequence similarity (100%) and phylogeny based on ClustalW alignments indicated that SC2b is a strain of *Bacillus* sp. (Fig. 11). The sequence obtained (1426 bp) was submitted in the NCBI databases under the accession number JX512223.1.

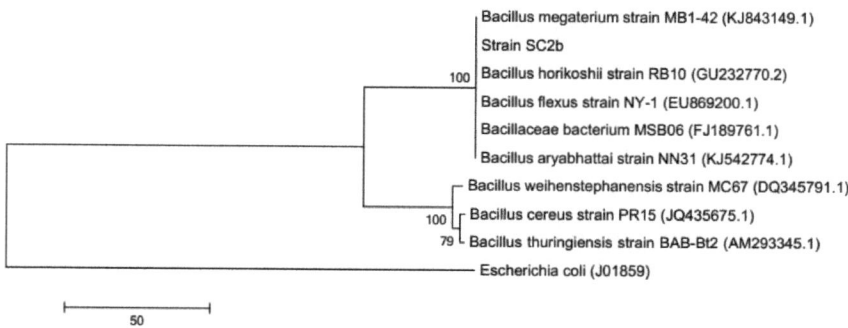

Figure 11 Phylogenetic tree showing the relationship of partial 16S rDNA gene sequences from metal mobilizing *Bacillus* sp. SC2b with other related sequences from *Bacillus*. *Escherichia coli* was used as the out-group. The value on each branch is the percentage of bootstrap replications supporting the branch.

3.2.   Plant growth-promoting features

Bacillus sp. SC2b showed positive for HCN production and was able to solubilize a significant amount of phosphate (56.6 mg/L) and produce IAA up to 64.8 mg/L after 5 days of incubation (Table 10). Data also demonstrated ACC deaminase activity at 25 µmαKB/mg/hr and displayed positive siderophore activity, as indicated by development of orange-colored zone on CAS agar plates after 5 days of growth. Under Fe-limiting conditions, strain SC2b produced 198.3 and 13.2

mg/L catechol and hydroxamate siderophore, respectively. Further, the effects of L-tryptophan concentration and pH on bacterial growth and IAA production by strain SC2b were also determined. As shown in Fig. 12A, SC2b produced the highest amount of IAA when cultured in LB broth amended with 2 mg/mL L-tryptophan, whereas at higher L-tryptophan concentrations (3, 4 and 5 mg/mL) an adverse effect on IAA production was observed. Strain SC2b was also able to grow in the medium over a wide range of initial pH ranging from 4 to 10; however, the highest synthesis of IAA by SC2b was obtained when cultivated in acidic media at pH 6 (Fig. 12B). Significant correlations were obtained between bacterial growth and IAA production at different L-tryptophan concentrations (Fig. 12A) and pH (Fig. 12B).

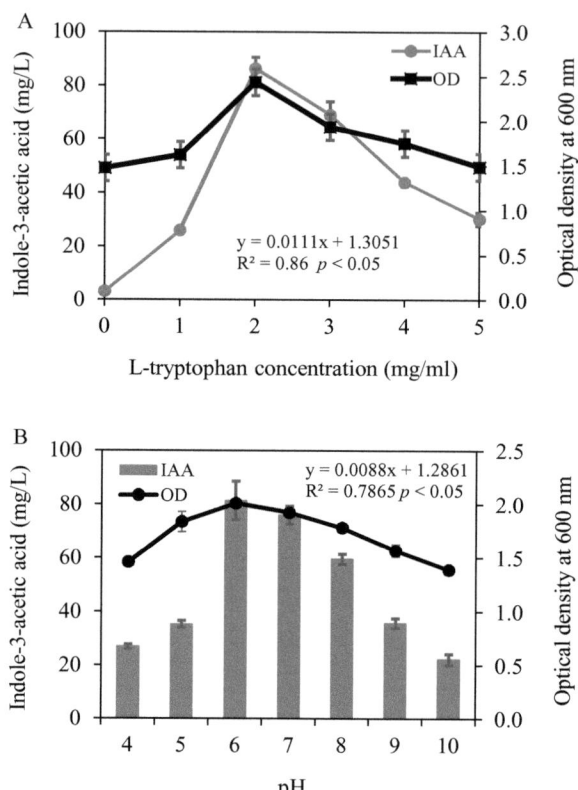

Figure 12 Impact of L-tryptophan concentration (A) and pH (B) on the growth of *Bacillus* sp. SC2b and its indole-3-acetic acid (IAA) production. Bars represent standard deviations of triplicates. Correlations between bacterial growth and IAA production at different L-tryptophan concentrations (A) and pH (B) were determined by linear regression analysis ($p < 0.05$).

Table 10 Biochemical characteristics of *Bacillus* sp. SC2b.

Characteristics	Parameter	Unit	*Bacillus* sp. SC2b
Metal resistance	Cd	mg/L	300
	Zn	mg/L	750
	Pb	mg/L	1400
Antibiotic resistance	Ampicillin	mm	4 (R)
	Tetracycline	mm	17 (S)
	Streptomycin	mm	18 (S)
	Chloramphenicol	mm	15 (I)
	Kanamycin	mm	17 (S)
Plant growth promoting feature	ACC deaminase production	$\mu m\alpha KB/mg/hr$	25 ±3.6
	P solubilization	mg/L	56.6 ±4.3
	IAA production	mg/L	64.8 ±2.0
	HCN production		+
	Siderophore CAS	cm	1.2 ±0.2
	Catechol	mg/L	198.3 ± 18.0
	Hydroxamate	mg/L	13.2 ±0.2

R, resistant (<10 mm); I, intermediate (10–15 mm); S, susceptible (>15 mm); ACC, 1-aminocyclopropane-1-carboxylate; α-KB, α-ketobutyrate; P, phosphate; IAA, indole-3-acetic acid; HCN, hydrogen cyanide; CAS, chrome azurol S; +, positive.

3.3. Metal biosorption and mobilization

As shown in Fig. 13, metal-resistant *Bacillus* sp. SC2b was capable of absorbing significant amounts of Cd, Pb and Zn. Maximal biosorption by *Bacillus* sp. SC2b was achieved after 8 hr incubation. Further incubation up to 10 hr did not enhance the extent of biosorption (data not shown). The highest amount of biosorption of heavy metals was observed with Zn, while the lowest was seen with Pb. At 50, 100 and 150 mg/L initial concentrations, the biosorption of Cd after 8 hr incubation was 3.8, 4.3, and 5.8 mg/g dry mass (Fig. 13A); Pb was 1.4, 4.2 and 3.1 mg/g dry mass (Fig. 13B); and Zn was 5.9, 7, and 10.8 mg/g dry mass, respectively (Fig. 13C). A batch assay was conducted to analyze metal mobilization potential of strain SC2b in multi-metal contaminated agricultural soil. Inoculation of strain SC2b for 10 days increased concentrations of water-extractable Cd, Pb and Zn in soil by 13.9, 4.5 and 3.8-fold, respectively, compared with non-inoculated control soil (Fig. 14).

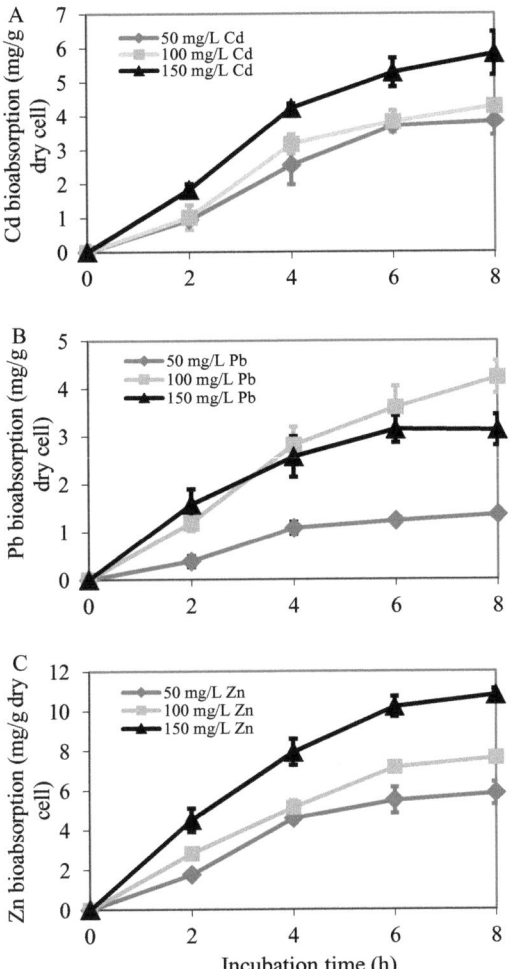

Figure 13 Biosorption of Cd (A), Pb (B) and Zn (C) on *Bacillus* sp. SC2b cells. Bars represent standard deviations of triplicates.

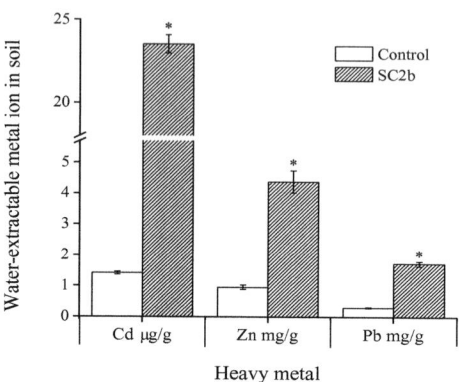

Figure 14 Effect of inoculation with *Bacillus* sp. SC2b on the mobilization of Cd, Pb and Zn in soil. Bars represent standard deviations of triplicates. An asterisk (*) denotes a value significantly greater than the corresponding control value according to Student's t-test ($p < 0.05$).

3.4.   Effects of inoculation of SC2b on the growth of Brassica napus

The PGP effects of *Bacillus* sp. SC2b were initially evaluated in the phytagar assay. It was found that inoculated *B. napus* exhibited 17% higher seed germination rate, 26% higher plant dry weight, 51% higher vigor index, 116 and 139% shoot and root elongation rates, respectively, compared with non-inoculated control treatments (Table 11).

Table 11 Plant growth promoting effects of *Bacillus* sp. SC2b in the microcosm assays.

Phytagar assay	Non-inoculated control	*Bacillus* sp. SC2b
% Germination	72 ±4 b	89 ±2 a
Shoot length (cm)	4.5 ±0.3 b	5.2 ±0.2 a
Root length (cm)	1.8 ±0.3 b	2.5 ±0.2 a
Shoot elongation rate (%) [1]	-	115.6 ±35
Root elongation rate (%) [2]	-	138.9 ±24
Vigor index [3]	454 ±36 b	685 ±42 a
Plant dry weight (g)	2.3 ±0.1 b	2.9 ±0.1 a
Pot experiment	Non-inoculated control	*Bacillus* sp. SC2b
Shoot fresh weight (g)	47.9 ±6.4 b	68.9 ±7.2 a
Root fresh weight (g)	0.7 ±0.0 b	2.2 ±0.6 a
Shoot dry biomass (g)	4.4 ±0.7 b	6.2 ±0.5 a
Root dry biomass (mg)	154.8 ±16.2 b	278.4 ±18.9 a
Chlorophyll a (mg/g fw)	1.3 ±0.2 b	1.9 ±0.3 a
Chlorophyll b (mg/g fw)	0.5 ±0.0 b	0.9 ±0.1 a
Total chlorophyll (mg/g fw)	1.9 ±0.3 b	3.0 ±0.4 a
Bacterial colonization (10^5 CFU/g)	nd	6.8 ±0.1

Values are means ±standard deviations of three samples. Data of rows indexed by the same letter are not significantly different between treatments according to Student's *t*-test ($p < 0.05$). fw, fresh weight; CFU, colony forming units; nd, not detected.

[1] Shoot elongation ratio (%) = Mean shoot length of tested plant / Mean shoot length of control x 100%

[2] Root elongation ratio (%) = Mean root length of tested plant / Mean root length of control x 100%

[3] Vigor index = germination (%) × seedling length (shoot length + root length)

3.5.   Effects of inoculation of SC2b on the growth and metal uptake of Sedum plumbizincicola

The influence of inoculation of strain SC2b on *S. plumbizincicola* growth and metal (Cd, Pb and Zn) uptake was determined in a pot experiment. Plants inoculated with strain SC2b exhibited a 46 and 42% increase in plant fresh and dry weight, respectively, compared with non-inoculated control (Table 11). In addition, SC2b inoculation also elevated Chl a, Chl b and total Chl content in leaves by 46, 80 and 58%, respectively. Inoculation of SC2b also increased concentrations of Cd and Zn in *S. plumbizincicola* by 15 and 13%, respectively over the non-inoculated controls (Fig. 15A and 15C). In contrast, inoculation of SC2b decreased concentrations of Pb in the plant tissues (Fig. 15B). The strain SC2b showed high level of colonization on rhizosphere of *S. plumbizincicola* (6.8 x 10^5 CFU/g) grown in multi-metal contaminated agricultural soil (Table 11).

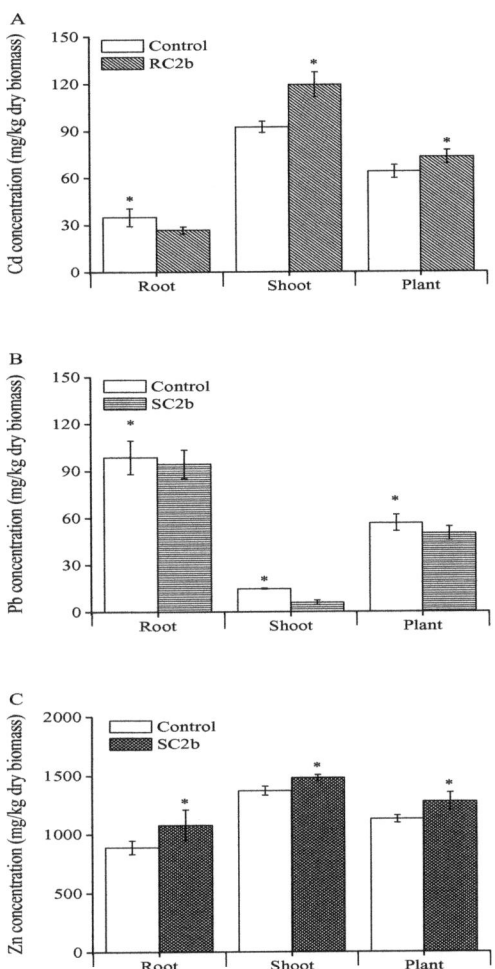

Figure 15 Effect of *Bacillus* sp. SC2b on the uptake of Cd (A), Pb (B) and Zn (C) by *Sedum plumbizincicola*. Bars represent standard deviations of triplicates. An asterisk (*) denotes a value significantly greater than the corresponding control value according to Student's t-test ($p < 0.05$).

4.   Discussion

The natural ability of plants in removal of heavy metals from polluted soils may be integrated and improved by metal-resistant PGPB, which are naturally present in metal polluted rhizosphere soils where they play important roles in plant growth and stress tolerance (Ma et al., 2011a; Ahemad and Kibret, 2014). In addition to improving plant growth, PGPB are themselves involved in metal immobilization, mobilization or transformation through various mechanisms including physical sequestration, exclusion, complexation and detoxification (Rajkumar et al., 2010). PGPB inoculation enhances plant growth as well as tolerance towards various heavy metals including Cd (Dell'Amico et al., 2008), nickel (Ni) (Ma et al., 2011b) and Cu (Zhang et al., 2011). In some cases, increased growth and tolerance against heavy metal stress observed in PGPB inoculated plants has been explained by plant beneficial traits of PGPB and/or reduced metal uptake by plants (Rajkumar et al., 2013). However, different PGPB-hyperaccumulator associations might provide different responses, and therefore, further research is required to understand whether PGPB inoculated plants and specific strains favor rhizoremediation of pollutants and through which mechanisms.

Although heavy metals exert inhibitory effects on microorganisms by displacing essential element ions or hindering functional groups, bacterial strains isolated from different habitats may exhibit different degrees of metal resistance, and those from metal polluted soils are usually more resistance (Ma et al., 2011a). Thus, bacterial strains isolated from metal polluted natural soils may be exploited for heavy metal bioremediation. Several bacterial strains isolated from metal polluted soils tolerant to Cd, Pb and Zn were previously reported (Becerra-Castro et al., 2012). Our results showed that *Bacillus* sp. SC2b was tolerant to high concentrations of Cd, Pb and Zn. This bacterial strain was isolated from a multi-metal polluted agricultural soil, where it may have evolved a strong resistance to heavy metals.

Various studies confirmed that metal-resistant bacteria possessing plant beneficial traits increase plant growth, nutrient uptake, metal tolerance and/or rhizoremediation process in polluted soils (Rajkumar et al., 2008; Ma et al., 2011b; Zhang et al., 2011). In this study, strain SC2b possessed multiple PGP traits such as production of IAA, siderophore, utilization of ACC, and solubilization of phosphate. Among the PGP traits, bacterial synthesis of IAA was found to help plants in production of longer roots for nutrient uptake under metal stress conditions (Golubev et al., 2011). Thus, the intent was to optimize production of IAA by supplementing *Bacillus* sp. SC2b with different level of L-tryptophan and growing the bacterial strain under various pH conditions.

As a precursor of IAA, the addition of L-tryptophan to bacterial cultures generally enhanced IAA biosynthesis (Costacurta and Venderleyden, 1995). In our study, IAA was produced in negligible quantities in L-tryptophan free medium. Addition of 2 mg/mL L-tryptophan of culture media resulted in an increase in production of IAA by this isolate, whereas higher concentrations of L-tryptophan (3, 4 and 5 mg/mL) showed negative effects on IAA production (Fig. 12A). The results indicate that bacterial IAA production was modulated by adding L-tryptophan and excessive L-tryptophan may result in the synthesis of IAA degrading/metabolizing enzymes such as IAA oxidase and peroxidase, and polyphenol oxidase (Datta and Basu, 2000). The regression analysis of bacterial growth and IAA production in medium indicated that conversion of L-tryptophan into IAA is closely dependent on growth and activity of the bacterial strain. The pH of the medium influenced the growth of *Bacillus* sp. SC2b and consequently the production of IAA, as reported by Acuña et al. (2011) in studies with other *Bacillus* spp.

The PGP effects of strain SC2b were further evaluated on *B. napus* using phytagar assay. Inoculation of strain SC2b exhibited a significant increase in germination rate, length and elongation rate of shoot and root, vigor index and dry weight of plants. Previous studies demonstrated the potential of rhizosphere bacteria, which possessed various PGP traits to promote the root elongation and growth of *Brassica juncea* and *B. napus* (Sheng and Xia, 2006; Ma et al., 2009). The enhanced growth response of plants induced by SC2b inoculation showed the ability of organisms to survive on the root and exhibit beneficial effects on the host plant growth. Considering such potential, a pot experiment was performed with an objective to assess usefulness of SC2b and *S. plumbizincicola* on rhizoremediation of metal polluted agricultural soils. Strain SC2b significantly elevated fresh and dry weight of *S. plumbizincicola* in metal polluted soils indicating that plants adapt to metal stress in soils more effectively with the help of rhizosphere bacteria.

The survival and colonization of bacteria in the rhizosphere are important aspects to evaluate the role of the inoculated PGPB in microbe-assisted phytoremediation of contaminated sites (Compant et al., 2010). Although the success of microbe assisted rhizoremediation process depends upon heavy metal uptake by plants, survival and activity of metal-resistant PGPB markedly influence the level of metal uptake by plants through metal mobilization or immobilization process (Ma et al., 2011a; 2015). The results of the present study also showed high level of colonization by strain SC2b, indicating the ability of this strain to survive and develop in the metal polluted rhizosphere of *S. plumbizincicola*. Further, the ability of SC2b in alleviating metal stress in *S. plumbizincicola* was also demonstrated by a significant rise in Chl a, Chl b and total Chl content in inoculated plants. The enhanced germination rate, biomass and related physiological parameters of plants produced by strain SC2b may be attributed to its PGP features, such as solubilization of phosphate and production of IAA, ACC deaminase and siderophores (Ma et al., 2011a; Ahemad and Kibret, 2014). Our results corroborate those of Dell'Amico et al. (2005), suggesting that the siderophore producing bacteria facilitate Fe uptake in plants through formation of mobile Fe–siderophore complexes thus favoring both chloroplast development and Chl biosynthesis (Rajkumar et al., 2010). In addition, strain SC2b exhibited a high degree of metal biosorption potential (Fig. 13). The capacity of bacterial adsorption of metals is generally dependent upon ionic radius of each metal (Karakagh et al., 2012). Data demonstrated that Zn (0.88 Å) having smaller ionic radius may be more rapidly complexed by bacteria compared to Cd (0.97 Å) and Pb (1.2 Å). Ma et al. (2011a) reported that PGPB inoculation decreased metal toxicity through biosorption and bioaccumulation mechanisms, thus exerting a protective effect on host plants against heavy metal toxicity and leading to higher plant growth and yields. Nevertheless, the mechanism underlying bacterial biosorption in protecting effect on plant against metal stress is poorly understood.

Accumulation of Cd, Pb and Zn in roots and shoots of *S. plumbizincicola* with or without PGPB inoculation was determined in the pot experiment. In general, shoot system accumulated significantly more Cd and Zn than root system, irrespective of inoculation treatment (Fig. 15). This might be attributed to the effective translocation of heavy metals from root to shoot system (Jiang et al., 2010). As shown in Fig. 15 (A and C), inoculation of strain SC2b significantly increased concentrations of Cd and Zn in plant system by 15 and 13%, respectively. These findings are in agreement with Rajkumar et al. (2008), who found that PGPB *Bacillus weihenstephanensis* SM3 inoculated onto *Helianthus annuus* enhanced Zn accumulation in root and shoot tissues by 22 and 35% respectively, compared with non-inoculated plants. The elevated concentrations of Cd and Zn in SC2b inoculated *S. plumbizincicola* corresponds to the effect of the bacterial strain on metal mobilization in soil (Fig. 14). Sessitsch et al. (2013) found that the presence of metal-resistant bacteria induced acidification of rhizosphere soils of plants by producing organic acids or siderophores, which enhance metal bioavailability around the root zone and thus facilitate plant metal uptake. In our study, inoculation with the metal-resistant bacterial strain SC2b significantly increased soil water-extractable metals Cd, Pb and Zn concentrations (Fig. 14), in accordance with previous finding of Rajkumar et al. (2008). This was probably attributed to acidification and chelation reactions in soil, which were initiated by solubilization of phosphate and production of catechol and hydroxamate siderophores. Further studies are needed to assess fate of metals in soil solution, as they leach from polluted agricultural soils and contribute to groundwater contamination.

Bacterial inoculation decreased Pb accumulation in root and shoot systems by 26 and 46%, respectively. Similar observations were also reported by Rajkumar et al. (2013) who found that inoculation of metal-resistant *B. megaterium* on surface-sterilized seeds of *B. juncea, Luffa cylindrica* and *Sorghum halepense* significantly reduced concentration of Ni in roots and shoots compared with non-inoculated plants. This situation was probably due to direct dilution of Ni concentration by increased plant biomass.

5.   Conclusions

Data demonstrated that inoculation of *Bacillus* sp. SC2b not only promoted biomass of plants, but also enhanced uptake of Cd and Zn in plant tissues especially in shoots. These beneficial effects produced by inoculation with *Bacillus* sp. SC2b, together with metal biosorption and biomobilization potential, indicate that metal-resistant PGPB possess potential to improve rhizoremediation efficiency of metal-contaminated soils.

Acknowledgements

Y. Ma thankfully acknowledges the Portuguese Foundation for Science and Technology (FCT) for awarding a post-doctoral research grant (SFRH/BPD/76028/2011).

References

Acuña, J. J., Jorquera1, M. A., Martínez, O. A., Menezes–Blackburn, D., Fernández, M. T., Marschner, P., Greiner, R., Mora, M. L., 2011. Indole acetic acid and phytase activity produced by rhizosphere bcilli as affected by pH and metals. J. Soil Sci. Plant Nutr. 11, 1– 12.

Ahemad, M., Kibret, M., 2014. Mechanisms and applications of plant growth promoting rhizobacteria: Current perspective. J. King Saud. Univ. Sci. 26, 1–20.

Altschul, S.F., Madden, T.L., Schaffer, A.A., Zhang, J., Zhang, Z., Miller, W., Lipman, D.J., 1997. Gapped BLAST and PSI-BLAST: A new generation of protein database search programs. Nucleic Acids Res. 25, 3389–3402.

Arnow, E., 1937. Colorimetric determination of the components of 3,4-dihydroxyphenylalanine-tyrosine mixtures. J. Biol. Chem. 118, 531–537.

Atkin, C.L., Neilands, J.B., Phaff, H.J., 1970. Rhodotorulic acid from species of *Leucosporidium*, *Rhodosporidium*, *Rhodotorula*, *Sporidiobolus*, and *Sporobolomyces*, and a new alanine-containing ferrichrome from *Cryptococcus melibiosum*. J. Bacteriol. 103, 722–733.

Bakker, A. W., Schippers, B., 1987. Microbial cyanide production in the rhizosphere in relation to potato yield reduction and *Pseudomonas* spp. mediated plant growth reduction. Soil Biol. Biochem. 19, 452–458.

Becerra-Castro, C., Monterrosob, C., Prieto-Fernández, A., Rodríguez-Lamas, L., Loureiro-Viñas, M., Acea, M. J., Kidd, P. S., 2012. Pseudometallophytes colonising Pb/Zn mine tailings: A description of the plant–microorganism–rhizosphere soil system and isolation of metal-tolerant bacteria. J. Hazard. Mater. 217–218, 350–359.

Belimov, A.A., Dodd, I.C., Hontzeas, N., Theobald, J.C., Safronova, V.I., Davies, W.J., 2009. Rhizosphere bacteria containing 1-aminocyclopropane-1-carboxylate deaminase increase yield of plants grown in drying soil via both local and systemic hormone signalling. New Phytologist 181, 413–423.

Berleth, T., Sachs, T., 2001. Plant morphogenesis: long distance coordination and local patterning. Curr. Opin. Plant Biol. 4, 57–62.

Biari, A., Gholami, A., Rahmani, H.A., 2008. Growth promotion and enhanced nutrient uptake of maize (*Zea mays* L.) by application of plant growth promoting rhizobacteria in arid region of Iran. J. Biol. Sci. 8, 1015–1020.

Bradford, M.M., 1976. A rapid and sensitive method for the quantitation of microgram quantities of protein utilizing the principle of protein-dye binding. Anal. Biochem. 72, 248–254.

Branco, R., Chung, A.P., Veríssimo, A., Morais, P.V., 2005. Impact of chromium contaminated wastewaters on the microbial community of a river. FEMS Microbiol. Ecol. 54, 35–46.

Bric, J.M., Bostock, R.M., Silversone, S.E., 1991. Rapid in situ assay for indole acetic acid production by bacteria immobilization on a nitrocellulose membrane. Appl. Environ. Microbiol. 57, 535–538.

Coelho. P., Costa, S., Silva, S., Walter, A., Ranville, J., Sousa, A. C., Costa, C., Coelho, M., García-Lestón, J., Pastorinho, M. R., Laffon, B., Pásaro, E., Harrington, C., Taylor, A., Teixeira, J. P., 2012. Metal(Loid) levels in biological matrices from human populations exposed to mining contamination—Panasqueira Mine (Portugal). J. Toxicol. Environ. Health A 75, 893–908.

Compant, S., Clément, C., Sessitsch, A., 2010. Plant growth-promoting bacteria in the rhizo- and endosphere of plants: Their role, colonization, mechanisms involved and prospects for utilization. *Soil Biol. Biochem.* 42, 669–678.

Costacurta, A., Venderleyden, J., 1995. Synthesis of phytohormones by plant associated bacteria. Crit. Rev. Microbiol. 21, 1–18.

Dahshan, H., Abd-Elall, A. M., Megahed, A. M., 2013. Trace metal levels in water, fish, and sediment from River Nile, Egypt: Potential health risks assessment. J. Toxicol. Environ. Health A 76, 1183–1187.

Datta, C., Basu, P.S., 2000. Indole acetic acid production by a rhizobium species from root nodules of a leguminous shrub, *Cajanus cajan*. Microbiol. Res. 155, 123–127.

Dell'Amico, E., Cavalca, L., Andreoni, V., 2005. Analysis of rhizobacterial communities in perennial Graminaceae from polluted water meadow soil, and screening of metal-resistant, potentially plant growth-promoting bacteria. FEMS Microbiol. Ecol. 52, 153–162.

Dell'Amico, E., Cavalca, L., Andreoni, V., 2008. Improvement of *Brassica napus* growth under cadmium stress by cadmium-resistant rhizobacteria. Soil Biol. Biochem. 40, 74–84.

Dworkin, M., Foster, J., 1958. Experiments with some microorganisms which utilize ethane and hydrogen. J. Bacteriol. 75, 592–601.

Fenton, T.W., Fenton, M., 1979. An improved procedure for the determination of chromic oxide in feed and feces. Can. J. Anim. Sci. 59, 631–634.

Figueiredo, N. L., Areias, A., Mendes, R., Canário, J., Duarte, A., Carvalho, C., 2014a. Mercury-resistant bacteria from salt marsh of Tagus Estuary: The influence of plants presence and mercury contamination levels. J. Toxicol. Environ. Health A. 77, 959–71.

Figueiredo, N. L., Canário, J., Duarte, A., Serralheiro, M. L., Carvalho, C., 2014b. Isolation and characterization of mercury-resistant bacteria from sediments of Tagus Estuary (Portugal): Implications for environmental and human health risk assessment. J. Toxicol. Environ. Health A. 77, 155–168.

Fiske, C.H., Subbarow, Y., 1925. A colorimetric determination of phosphorus. J. Biol. Chem. 66, 375–400.

Freitas, H., Prasad, M.N.V., Pratas, J., 2004. Analysis of serpentinophytes from northeast of Portugal for trace metal accumulation relevance to the management of mine environment. Chemosphere 54, 1625–1642.

Gaonkar, T., Bhosle, S., 2013. Effect of metals on a siderophore producing bacterial isolate and its implications on microbial assisted bioremediation of metal contaminated soils. Chemosphere 93, 1835–1843.

Ghosh, P., Rathinasabapathi, B., Ma, L.Q., 2011. Arsenic-resistant bacteria solubilized arsenic in the growth media and increased growth of arsenic hyperaccumulator *Pteris vittata* L. Bioresour. Technol. 102, 8756–8761.

Giller, K.E., Witter E., McGrath S.P., 1998. Toxicity of heavy metals to microorganisms and microbial processes in agricultural soils: A review. Soil Biol. Biochem. 30, 1389–1414.

Glick, B.R., 2003. Phytoremediation: synergistic use of plants and bacteria to clean up the environment. Biotechnol. Adv. 21, 383–393.

Glick, B.R., Cheng, Z., Czarny, J., Duan, J., 2007. Promotion of plant growth by ACC deaminase-producing soil bacteria. Eur. J. Plant Pathol. 119, 329–339.

Glick, B. R., 2010. Using soil bacteria to facilitate phytoremediation. Biotechnol. Adv. 28, 367–374.

Golubev, S. N., Muratova, A. Y., Wittenmayer, L., Bondarenkova, A. D., Hirche, F., Matora, L. Y., Merbach, W., Turkovskaya, O. V., 2011. Rhizosphere indole-3-acetic acid as a mediator in the *Sorghum bicolor*-phenanthrene-*Sinorhizobium meliloti* interactions. Plant Physiol. Biochem. 49, 600–608.

Gordon, S.A., Weber, R.P., 1951. Colorimetric estimation of indoleacetic acid. Plant Physiol. 26, 192–195.

He, H.D., Ye, Z.H., Yang, D.J., Yan, J.L., Xiao, L., Zhong, T., Yuan, M., Cai, X.D., Fang, Z.Q., Jing, Y.X., 2013. Characterization of endophytic *Rahnella* sp. JN6 from *Polygonum pubescens* and its potential in promoting growth and Cd, Pb, Zn uptake by *Brassica napus*. Chemosphere 90, 1960–1965.

Holden, T., 1989. How to select hazardous waste treatment technologies for soils and sludges: alternative innovative and emerging technologies. Park Ridge, USA: Noyes Data Corporation.

Honma, M., Shimomura, T., 1978. Metabolism of 1-aminocyclopropane-1-carboxylic acid. Agr. Biol. Chem. 42, 1825–1831.

Inskeep, W. P., Bloom, P. R., 1985. Extinction coefficients of chlorophyll a and b in N,N-dimethylformamide and 80% acetone. Plant Physiol. 77, 483–485.

Jeong, S., Moon, H.S., Nam, K., Kim, J.Y., Kim, T.S., 2012. Application of phosphate-solubilizing bacteria for enhancing bioavailability and phytoextraction of cadmium (Cd) from polluted soil. Chemosphere 88, 204–210.

Jiang, C.Y., Sheng, X.F., Qian, M., Wang, Q.Y., 2008. Isolation and characterization of a heavy metal-resistant *Burkholderia* sp. from heavy metal-contaminated paddy field soil and its potential in promoting plant growth and heavy metal accumulation in metal polluted soil. Chemosphere 72, 157–164.

Jiang, J.P., Wu, L.H., Li, N., Luo, Y.M., Liu, L., Zhao, Q.G., Zhang, L., Christie, P., 2010. Effects of multiple heavy metal contamination and repeated phytoextraction by *Sedum plumbizincicola* on soil microbial properties. Eur. J. Soil Biol. 46, 18–26.

Kachenko, A.G., Singh, B., 2006. Heavy metals contamination in vegetables grown in urban and metal smelter contaminated sites in Australia. Water Air Soil Pollut. 169, 101–123.

Kamaludeen, S. P., Ramasamy, K., 2008. Rhizoremediation of metals: harnessing microbial communities. Indian J. Microbiol. 48, 80–88.

Kane, M.D., Poulsen, L.K., Stahl, D.A., 1993. Monitoring the enrichment and isolation of sulfate-reducing bacteria by using oligonucleotide hybridization probes designed from environmentally derived 16S rRNA sequences. Appl. Environ. Microbiol. 59, 682–686.

Karakagh, R. M., Chorom, M., Motamedi, H., Kalkhajeh, Y. K., Oustan, S. 2012. Biosorption of Cd and Ni by inactivated bacteria isolated from agricultural soil treated with sewage sludge. Ecohydrol. Hydrobiol. 12, 191–198.

Kuiper, I., Lagendijk, E. L., Bloemberg, G. V., Lugtenberg, B. J., 2004. Rhizoremediation: A beneficial plant-microbe interaction. Mol. Plant Microbe Interact. 17, 6–15.

Khamna, S., Yokota, A., Peberdy, J.F., Lumyong, S., 2010. Indole-3-acetic acid production by *Streptomyces* sp. isolated from some Thai medicinal plant rhizosphere soils. EurAsia. J. BioSci. 4, 23–32.

Lane, D.J., 1991. 16S/23S rRNA sequencing in nucleic acid techniques in bacterial systematics. In: Stackebrandt, E., Goodfellow, M. (Eds.) Wiley, New York, pp. 115–175.

Legault, G.S., Lerat, S., Nicolas, P., Beaulieu, C., 2011. Tryptophan regulates thaxtomin A and indole-3-acetic acid production in *Streptomyces scabiei* and modifies its interactions with radish seedlings. Phytopathology 101, 1045–1051.

Li, M. S., 2005. Ecological restoration of mineland with particular reference to the metalliferous mine wasteland in China: A review of research and practice. Sci. Total Environ. 357, 38–53.

Li, Y., Wang, Y.B., Gou, X., Su, Y.B., Wang, G., 2006. Risk assessment of heavy metals in soils and vegetables around non-ferrous metals mining and smelting sties, Baiyin, China. J. Environ. Sci. (China) 18, 1124–1134.

Li, Z. Y., Ma, Z. W., van der Kuijp, T. J., Yuan, Z. W., Huang, L., 2014. A review of soil heavy metal pollution from mines in China: Pollution and health risk assessment. Sci. Total Environ. 468–469, 843–853.

Lowry, O.H., Rosebrough, N.J., Farr, A.L., Randall, R.J., 1951. Protein measurement with the folin phenol reagent. J. Biol. Chem. 193, 265–275.

Ma, Y., Rajkumar, M., Freitas, H., 2009. Improvement of plant growth and nickel uptake by nickel resistant-plant growth promoting bacteria. J. Hazard. Mater. 166, 1154–1161.

Ma, Y., Rajkumar, M., Vicente, J., Freitas, H., 2010. Inoculation of Ni-resistant plant growth promoting bacterium *Psychrobacter* sp. strain SRS8 for the improvement of nickel phytoextraction by energy crops. Int. J. Phytoremediat. 13, 126–139.

Ma, Y., Prasad, M.N.V., Rajkumar, M., Freitas, H., 2011a. Plant growth promoting rhizobacteria and endophytes accelerate phytoremediation of metalliferous soils. Biotechnol. Adv. 29, 248–258.

Ma, Y., Rajkumar, M., Luo, Y.M., Freitas, H., 2011b. Inoculation of endophytic bacteria on host and non-host plants – effects on plant growth and Ni uptake. J. Hazard. Mater. 196, 230–237.

Ma, Y., Rajkumar, M., Luo, Y., Freitas, H., 2013. Phytoextraction of heavy metal polluted soils using *Sedum plumbizincicola* inoculated with metal mobilizing *Phyllobacterium myrsinacearum* RC6b. Chemosphere 93, 1386–1392.

Ma, Y., Rajkumar, M., Rocha, I., Oliveira, R. S., Freitas, H., 2015. Serpentine bacteria influence metal translocation and bioconcentration of *Brassica juncea* and *Ricinus communis* grown in multi-metal polluted soils. Front. Plant Sci. 5, 757.

Martñez-Alcalá, I., Clemente, R., Bernal, M.P., 2009. Metal availability and chemical properties in the rhizosphere of *Lupinus albus* L. growing in a high-metal calcareous soil. Water Air Soil Pollut. 201, 283–293.

Mateos, P.F., Jimenez-Zurdo, J.I., Chen, J., Squartini, A.S., Haack, S.K., Martinez-Molina, E., Hubbell, D.H., Dazzo, F.B., 1992. Cell-associated pectinolytic and cellulolytic enzymes in *Rhizobium leguminosarum* bv. *Trifolii*. Appl. Environ. Microbiol. 58, 1816–1822.

McGrath, S.P., Lombi, E., Gray, C.W., Caille, N., Dunham, S.J., Zhao, F.J., 2006. Field evaluation of Cd and Zn phytoextraction potential by the hyperaccumulators *Thlaspi caerulescens* and *Arabidopsis halleri*. Environ. Pollut. 141, 115–25.

MEPPRC (Ministry of Environmental Protection of the People's Republic of China), 2006. Report on the State of the Environment in China. Beijing, China (in Chinese).

Mishra, P. K., Mishra, S., Bisht, S. C., Selvakumar, G., Kundu, S., Bisht, J. K., Gupta, H. S., 2009. Isolation, molecular characterization and growth-promotion activities of a cold tolerant bacterium *Pseudomonas* sp. NARs9 (MTCC9002) from the Indian Himalayas. Biol. Res. 42, 305–313.

Muleta, D., Assefa, F., Börjesson, E., Granhall, U., 2013. Phosphate-solubilising rhizobacteria associated with *Coffea arabica* L. in natural coffee forests of southwestern Ethiopia. J. Saudi Soc. Agr. Sci. 12, 73–84.

Murphy, J., Riley, J.P., 1962. A modified single solution method for the determination of phosphorus in natural waters. Anal. Chim. Acta. 12, 31–36.

Nies, D.H., 2003. Efflux-mediated heavy metal resistance in prokaryotes. FEMS Microbiol. Rev. 27, 313–339.

NSPRC (National Standards of the People's Republic of China), 1995. Standards for soil environmental quality: GB15618–1995.

Oves, M., Khan, M.S., Zaidi, A., 2013. Biosorption of heavy metals by *Bacillus thuringiensis* strain OSM29 originating from industrial effluent contaminated north Indian soil. Saudi J. Biol. Sci. 20, 121-129.

Pal, A., Dutta, S., Mukherjee, P.K., Paul, A.K., 2005. Occurrence of heavy metal-resistance in microflora from serpentine soil of Andaman. J. Basic Microbiol. 45, 207–218.

Park, J.H., Bolan, N., Mallavarapu, M., Naidu, R., 2011. Isolation of phosphate solubilizing bacteria and their potential for lead immobilization in soil. J. Hazard. Mater. 185, 829–836.

Park, J.H., Bolan, N., 2013. Lead immobilization and bioavailability in microbial and root interface. J. Hazard. Mater. http://dx.doi.org/10.1016/j.jhazmat.2013.02.010.

Patil, V., 2011. Production of indole acetic acid by *Azotobacter* sp. Recent Res. Sci. Tech. 3, 14–16.

Patten, C., Glick, B., 2002. Role of *Pseudomonas putida* indole acetic acid in development of the host plant root system. Appl. Environ. Microbiol. 68, 3795–3801.

Prapagdee, B., Chanprasert, M., Mongkolsuk, S., 2013. Bioaugmentation with cadmium-resistant plant growth-promoting rhizobacteria to assist cadmium phytoextraction by *Helianthus annuus*. Chemosphere. http://dx.doi.org/10.1016/j.chemosphere.2013.01.082.

Rajkumar, M., Ma, Y., Freitas, H., 2008. Characterization of metal-resistant plant-growth promoting *Bacillus weihenstephanensis* isolated from serpentine soil in Portugal. J. Basic Microbiol. 48, 1–9.

Rajkumar, M., Ae, N., Freitas, H., 2009. Endophytic bacteria and their potential to enhance heavy metal phytoextraction. Chemosphere 77, 153–160.

Rajkumar, M., Sandhya, S., Prasad, M.N.V., Freitas, H., 2012. Perspectives of plant-associated microbes in heavy metal phytoremediation. Biotechnol. Adv. 30, 1562–1574.

Rajkumar, M., Ma, Y., Freitas, H., 2013a. Improvement of Ni phytostabilization by inoculation of Ni resistant *Bacillus megaterium* SR28C. J. Environ. Manag. 128, 973–980.

Rajkumar, M., Prasad, M.N.V., Sandhya, S., Freitas, H., 2013b. Climate change driven plant-metal-microbe interactions. Environ. Int. 53, 74–86.

Rashid, M., Khalil, S., Ayub, N., Alam, S., Latif, F., 2004. Organic acids production and phosphate solubilization by phosphate solubilizing microorganisms (PSM) under in vitro conditions. Pakistan J. Biol. Sci. 7, 187–196.

Reeves, R.D., Baker, A.J.M., 2000. Metal-accumulating plants. In: Raskin, I., Ensley, B.D. (Eds.) Phytoremediation of toxic metals: using plants to clean up the environment. Wiley, New York, pp. 193–229.

Sanita di Toppi, L., Gabbrielli, R., 1999. Response to cadmium in higher plants. Environ. Exp. Bot. 41, 105–130.

Schwyn, B., Neilands, J.B., 1987. Universal chemical assay for the detection and determination of siderophores. Anal. Biochem. 160, 47–56.

Sessitsch, A., Kuffner, M., Kidd, P., Vangronsveld, J., Wenzel, W.W., Fallmann, K., Puschenreiter, M., 2013. The role of plant-associated bacteria in the mobilization and phytoextraction of trace elements in contaminated soils. Soil Biol. Biochem. 60, 182–194.

Sheng, X. F., Xia, J. J., 2006. Improvement of rape (*Brassica napus*) plant growth and cadmium uptake by cadmium-resistant bacteria. Chemosphere 64, 1036–1042.

Sheng, X.F., He, L.Y., Wang, Q.Y., Ye, H.S., Jiang, C., 2008. Effects of inoculation of biosurfactant producing *Bacillus* sp. J119 on plant growth and cadmium uptake in a cadmium amended soil. J. Hazard. Mater. 155, 17–22.

Srivastava, S., Verma, P.C., Chaudhry, V., Singh, N., Abhilash, P.C., Kumar, K.V., Sharma, N., Singh, N., 2013. Influence of inoculation of arsenic-resistant *Staphylococcus arlettae* on growth and arsenic uptake in *Brassica juncea* (L.) Czern. Var. R-46. J. Hazard. Mater. 262, 1039–1047.

Sundara-Rao, W.V.B., Sinha, M.K., 1963. Phosphate dissolving microorganisms in the soil and rhizosphere. Indian J. Agr. Sci. 33, 272–278.

Verma, S.C., Ladha, J.K., Tripathi, A.K., 2001. Evaluation of plant growth promoting and colonization ability of endophytic diazotrophs from deep water rice. J. Biotechnol. 91, 127–141.

Visioli, G., D'Egidio, S., Vamerali, T., Mattarozzi, M., Sanangelantoni, A.M., 2014. Culturable endophytic bacteria enhance Ni translocation in the hyperaccumulator *Noccaea caerulescens*. Chemosphere 117, 538–544.

Wan, Y., Luo, S.L., Chen, J.L., Xiao, X., Chen, L., Zeng, G.M., Liu, C.B., He, Y.J., 2012. Effect of endophyte-infection on growth parameters and Cd-induced phytotoxicity of Cd-hyperaccumulator *Solanum nigrum* L. Chemosphere 89, 743–750.

Wei, G.H., Fan, L.M., Zhu, W.F., Fu, Y.Y., Yu, J.F., Tang, M., 2009. Isolation and characterization of the heavy metal resistant bacteria CCNWRS33-2 isolated from root nodule of *Lespedeza cuneata* in gold mine tailings in China. J. Hazard. Mater. 162, 50–56.

Wenzel, W. W., 2009. Rhizosphere processes and management in plant-assisted bioremediation (phytoremediation) of soils. Plant Soil 321, 385–408.

Wu, L.H., Li, N., Luo, Y.M., 2008. Phytoextraction of heavy metal contaminated soil by *Sedum plumbizincicola* under different agronomic strategies. Proc. 5th Int. Phytotech. Conf., Nanjing, China, pp. 49–50.

Wu, L. H., Liu, Y. J., Zhou, S. B., Guo, F. G., Bi, D., Guo, X. H., Baker, A. J. M., Smith, J. A. C., Luo, Y. M., 2013. *Sedum plumbizincicola* X.H. Guo et S.B. Zhou ex L.H. Wu (Crassulaceae): a new species from Zhejiang Province, China. Plant Systemat. Evol. 299, 487–498.

Yilmaz, E.I., 2003. Metal tolerance and biosorption capacity of *Bacillus circulans* strain EB1. Microbiology 154, 409–415.

Zhang, Y.F., He, L.Y., Chen, Z.J., Wang, Q.Y., Qian, M., Sheng, X.F., 2011. Characterization of ACC deaminase-producing endophytic bacteria isolated from copper-tolerant plants and their potential in promoting the growth and copper accumulation of *Brassica napus*. Chemosphere 1, 57–62.

Zhu, L.J., Guan, D.X., Luo, J., Rathinasabapathi, B., Ma, L.Q., 2014. Characterization of arsenic-resistant endophytic bacteria from hyperaccumulators *Pteris vittata* and *Pteris multifida*. Chemosphere 113, 9–16.

Zhuang, P., McBride, M.B., Xia, H., Li, N., Li, Z., 2009. Health risk from heavy metals via consumption of food crops in the vicinity of Dabaoshan mine, South China. Sci. Total Environ. 407, 1551–1561.

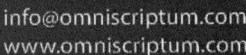

Printed by Books on Demand GmbH, Norderstedt / Germany